CHAMPION MINES

OUR FATHERS' MINES

A HISTORY OF THE

CHAMPION MINE

AND THE

TAIT'S GAP MINE

IN

ONEONTA, ALABAMA

The Lives and Times of the Miners and Their Families

WRITTEN AND COMPILED BY AULDEN WOODARD AND VAN GUNTER

Champion Mine - 1929

Blount County Memorial Museum
HISTORY SERIES

CHAMPION MINES

OUR FATHERS' MINES

Copyright 2011

WRITTEN AND COMPILED BY AULDEN WOODARD AND VAN GUNTER

All rights reserved.

Printed in the United States of America. No part of this book may be used or reproduced in any manner whatsoever without written permission except in the case of brief quotations embodied in critical articles and reviews.

Fifth Estate Publishing
Blountsville, AL 35031

First Edition

Cover Designed by An Quigley

Printed on acid-free paper

Library of Congress Control No: 2011941873

ISBN: 9781936533183

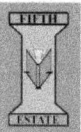

Fifth Estate, 2011

Champion Mines

Map of Champion & Tait's Gap

Map Showing Locations of Champion and Tait's Gap

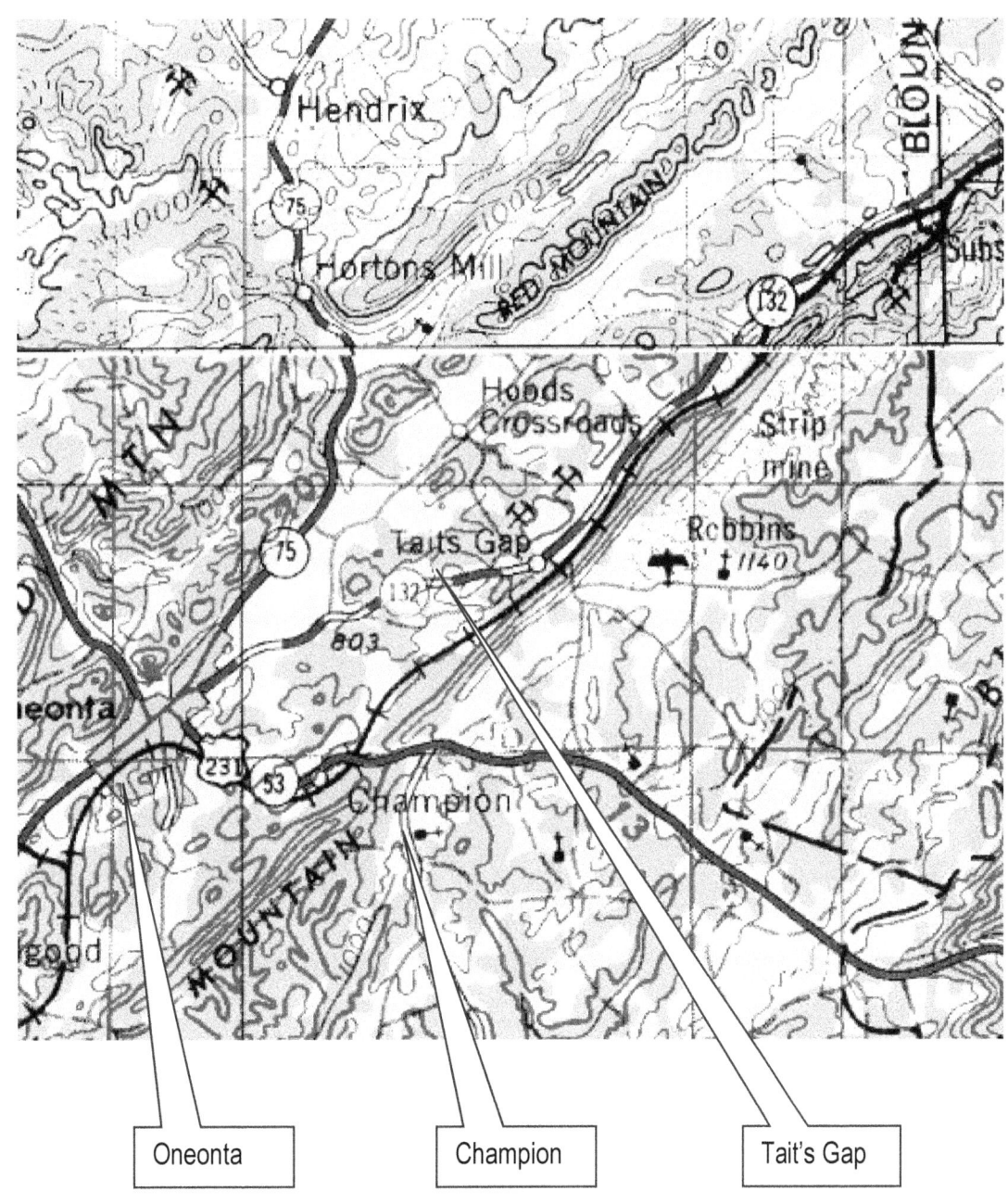

Preface

This book is a compilation of information and stories about the lives and times of the miners of Champion and Tait's Gap mines and their families. It was spearheaded by Aulden Woodard, Van Gunter, and Amy Rhudy and edited by Carol Lewey. Stories were submitted by the miners and their families listed below. Some of the material may be duplicated in another part of the book in order to complete a story. A very special thanks goes out to them for their help, without which this book could not have been written.

Champion Mine
Emma Linder
Quinton Lybrand
Krissy Taylor
Verlon Tipton
Marvin Tuck

Tait's Gap Mine
Peggy Bird
Fred Cornelius
Alfred Davenport
Jerome Parker
Margretta Smith

Champion Mines

FORWARD

Photo by Aulden Woodard, May 5, 1997   Emma Vandegrift Linder

This book pays tribute to Emma Vandegrift Linder, 1914-1999, Museum Curator and the daughter of Edgar N. Vandegrift, Superintendent of the Tait's Gap and Champion Mines from 1920-1965.

A charter member of the Blount County Historical Society, Mrs. Linder spearheaded the compilation and publication of <u>The History of Blount County</u> and disseminated information about county families through her genealogy column, "Lineage and Letters;" she also was instrumental in establishing the Warrior River's chapter of the American Revolution, was an active member of the Worthwhile Club, Blount County-Oneonta Chamber of Commerce, ARC of Blount County, and member of the Blount County Sportsman and Conservation Club.

Mrs. Linder was the recipient of numerous honors and awards, including the Governor's Conservation Award in 1963 and Gadsden Woman of the Year Award in 1955, as well as an exemplar of extraordinary leadership ability and achievement as a member of the Blount County Park Board, Oneonta and Amaryllis Garden Clubs, and Tait's Gap Home Demonstration Club.

From the: REPRESENTATIVES OF THE LEGISLATURE OF ALABAMA'S RESOLUTION OF CONDOLENCE, ON 25 MAY, 1999.

Champion Mines

TABLE OF CONTENTS

Map of Champion & Tait's Gap 32
Preface 4
Forward 5
   Emma Vandegrift Linder 5
Introduction 11
   MARKER ERECTED 1989 BY ALABAMA HISTORICAL ASSOCIATION 12
Chapter One 14
   MY FATHER'S MINES 14
Chapter Two 20
   OUR FATHERS' MINES 20
     Historical Champion 20
     Spout Spring Gap 22
Blount County Soil Map 26
   BLOUNTY COUNTY SOIL MAP 26
   CHAMPION MINERS HONOR ROLL 28
Chapter Three 28
   SUMMER TIME 29
Layout of Houses at Champion Mine – 1930s 31
Chapter Four 31
   CHAMP OF CHAMPION 32
Chapter Five 34
   A KID'S STORY 34
     Saturday Movie Memories 36
     Halloween 1941 Brings Heartache as School Burns 36
     Our Champion School Teachers 38
     Our Oneonta Teachers 39
Chapter Six 40
   THE BIRMINGHAM MINERAL RAILROAD 40
     Champion - Railroad Era Ends 44
     The New Champion Village Today 44
Chapter Seven 46
   THE SUPERINTENDENT 46
Chapter Eight 49
   FATHER AND GRANDFATHER 49
   Grady Woodard and Monroe Murphree, Miners 49
     Accident! 49
     New Housing 50
Chapter Nine 53
   MINER VERLON TIPTON, "MR. CHAMPION" 53
     Father's Accident 53
     Family and Camp Life 53
     Mines Close and Reopen 54
     Dam Breaks 55
Chapter Ten 57
   ARTHUR TIDWELL – LEADER AND SYRUP MAKER 57
     The Big Scare 58
Chapter Eleven 60

Champion Mines

WILEY GURLEY, STEAM SHOVEL OPERATOR	60
Steam Shovel	61

Chapter Twelve ... 64
  THOMAS RUSSELL (TOM) GREEN, MINER ... 64

Chapter Thirteen ... 65
  TAIT'S GAP MINES ... 65
  TAIT'S GAP MINERS HONOR ROLL - 1920s - 1950s ... 66
    Tait's Gap Map - 1930s ... 67

Chapter Fourteen ... 68
  AURELIA MAE ALVERSON VANDEGRIFT ... 68
  TAIT'S GAP MINERS' HERO ... 68

Chapter Fifteen ... 71
  TAIT'S GAP LIVES AND TIMES ... 71
    Clarence Moody Killed ... 71
    The Tait's Gap School ... 76
  WILLIE FRANKS AND HARLEY SPRADLIN - MINERS ... 77
  WILLIE A. FRANKS - MINER ... 79
    Accidents ... 79
    Tait's Gap mine accident in June, 1925 ... 80
    Accident ... 80
  Master Honor Roll ... 81

Brief History ... 84
  BRIEF HISTORY OF CHAMPION MINES ... 84

Washer Complex – left side ... 84

Section One ... 85
  PIONEER PERIOD ... 85
    War Between the States ... 86
    Rousseau's Raid Crosses Champion ... 86
    Rousseau Takes Alaska ... 87
    After the War ... 87
    Pig Iron - A New Process ... 88
    Champion Mines - The Beginning ... 89
    The Million Dollar Deal ... 89

Section Two ... 91
  STEAM POWER ERA ... 91
    Oneonta Incorporated ... 91
    Chepultepec - Too Difficult to Pronounce and Spell ... 92
    County Seat Moved ... 92
    T.C.I. Purchases Debardeleben's Holdings ... 93
    Champion Production ... 94
    Railroad Extended to Altoona ... 94

Section three ... 96
  THE SHOOK BROTHERS TAKE CHARGE ... 96
    Paschal Green Shook ... 96
    James Warner Shook ... 97
    Shook & Fletcher Gets Involved ... 97
    Ed Vandegrift Becomes Superintendent ... 98
    Riding the Train with My Grandmother ... 99
    Tait's Gap School ... 100
    Morrison Family ... 100
    Tait's Gap Mine ... 102

- Dinkeys 103
- Engine No. 22 104
- Charcoal Blast Furnaces 107
- Brown Ore 107
- Developing Tait's Gap 108
- Tornado! 110
- Dam Collapses 111
- Living at Tait's Gap 112
- Shook & Fletcher Lease Champion 114
- DIARY OF EDGAR NEWTON VANDEGRIFT – 1928 114
  - January 1928 115
  - February 1928 117
  - March 1928 118
  - April 1928 119
  - May 1928 120
  - June 1928 121
  - July 1928 122
  - August 1928 122
  - September 1928 123
  - October 1928 124
  - November 1928 125
  - December 1928 126

Section Four 128
- THE GREAT DEPRESSION YEARS 129
  - Mining and Living during the Depression 129
  - The Pink Slip 131
  - Days of the Depression 133
  - 1932 133
  - 1933 135
  - 1934 136
  - Miners Listed by Crew - June 1934 - In Vandegrift's Tait's Gap Time Book 142

Section Five 144
- FILES OF PASCHAL G. SHOOK 144
  - TCI Upheaval 145
  - Steam Shovel for Sale 146
  - Power Problems 147
  - Water Problems 147
  - Euclids 149
  - Automobile Accident 149
  - Pig Iron - Origin of the Term 149
  - Gypsum Crystal Found 150

Section Five 151
- THE FINAL MINING 151
  - Diesel Power and Heavy Media Plant 151
  - Tait's Gap Miners on Payroll, June 11, 1952 - In Vandegrift's Tait's Gap Time Book 152
  - Tait's Gap Mine Shipments of Ore and Tailings, August 1951 to September 1952 .. 153
  - Tait's Gap Miners on Payroll 154
  - June, 1956 to June 12, 1957 (working 60-hr weeks) 154
  - Tait's Gap Mine Shipments of Ore Tailings and Gravel, June 1956 to June 1957 ... 155
  - Tait's Gap Miners on Payroll. May 14, 1958- June 28, 1959 (60-hr Workweek) 156
  - Champion Mines Close for the Last Time 157
  - My Summer Jobs at Tait's Gap and Champion - 1961-1962 157

Champion Mines

  Second Summer 160
  Radio Hill 1961-1962 162
  Accidents 162
  Euclid Overturned 163
  Miners with whom I (Van Gunter) Worked - Summers - 1961 & 1962 163
Glossary 166
 Glossary of Ore Mining Terms 166
  References 168
 Memorabilia - Articles - Photographs 169
 From the Museum's Artifacts Files 169
Map - Blount County 170
clipping:. The Blount County iron industry? 171
Iron Industry clipping (continued) 172
Champion Miners Society Registration Form 2004 – 2010 173
Clipping: 174
L&N helped start Oneonta in 1888 174
Growth in Mining Areas - Article XIII (Louisville and Nashville Railroad: 1850 - 1963 by Kincaid A. Herr - 1964 175
Growth in Mining Areas - Article XIII (Louisville and Nashville Railroad: 1850 - 1963 by Kincaid A. Herr - 1964 176
Continued 176
Growth in Mining Areas - Article XIII (Louisville and Nashville Railroad: 1850 - 1963 by Kincaid A. Herr - 1964 177
Continued 177
Growth in Mining Areas - Article XIII (Louisville and Nashville Railroad: 1850 - 1963 by Kincaid A. Herr - 1964 178
Growth in Mining Areas - Article XIII (Louisville and Nashville Railroad: 1850 - 1963 by Kincaid A. Herr – 1964 Continued 179
Clippings(contonued) 183
Post Cards 184
Mines Ad - Tait's Gap Coal 186
Map ....Champion Map Camps – 1935 - In 3 page parts.
 187
Campion Map. (continued) 188
Draft - Letter from David T. Palmer to Dr. Payne, 7 November, 1996. *Blount County: A Geological Survey-* 190
Palmer Letter (continued) 193
Geological Survey of Alabama 195
Brown Ore Production by Year 196
letter from Shook & Fletcher 197
EChampion Mines Historecal Marker to be Dedicated Sunday 198
Tait's Gap School – October 2011 – with new entray/porch 201
Mr. Dowd, Washer Foreman, and his car – Champion Mines 1929 201

9

Champion Mines

<u>Favorite Past Time – Sitting atop Champion's Railroad Station Sign</u> 202
<u>Champion – Pump Pond 1928 – sitting on intake housing in pond</u> ... 202
<u>Equipment Service Checklist p. 2</u> 204
<u>Shipping Ticket</u> 205
<u>Timesheet</u> 205
<u>Flowchart of Work at the Mines</u> 206
<u>ABOUT THE WRITERS   AND THE MUSEUM</u> 208
   <u>Aulden Woodard</u> 208
   <u>Van Gunter</u> 208
   <u>Amy Rhudy</u> 211

INTRODUCTION

By Auden Woodard

This book tells the history of Champion, a mining camp community near Oneonta, Alabama, and the lives and times of its miners and their families. It shows how they struggled to live on only seven months of work a year and how their school and church helped them during the hard times in the southern Appalachian mountain camp. It recounts growing up in an isolated place with one tree swing and a company commissary store. It describes events of the camp and the railroad running through it. This book is a collection of history, museum material, and information from the miners and their descendants. The book is about families, fathers, sons, grandfathers, fathers-in-law, and grandsons who worked the mines. Some material was used from a book manuscript, The Southern Passage, by Grady Aulden Woodard.

Method of Mining at the Champion and Tait's Gap Mines

The iron ore mining industry uses the above-ground "strip" mining method to mine ore. In this method, the earth is stripped down to the level needed to get access to the pocket of ore. As usual, state-of-the-art equipment - steam shovels, drag lines, Dinkeys, and dump trucks - were commonly used. With the ore pocket exposed and using the same equipment, the red iron-ore-rich material is then hauled to a washer which separates the tailings from the ore using water. Rocks are removed and the ore is graded on size and quality for market.

The problem with this method was that it left a huge wasteland called "the cuts" as well as waste mud storage ponds from the washing process. The land was not normally reclaimed for use. Recent government environment regulations now enforce land reclamation. The coal industry uses the "strip" practices and caused this to get attention. Champion and Tait's Gap mines lands were not reclaimed.

Champion Mines

MARKER ERECTED 1989 BY ALABAMA HISTORICAL ASSOCIATION

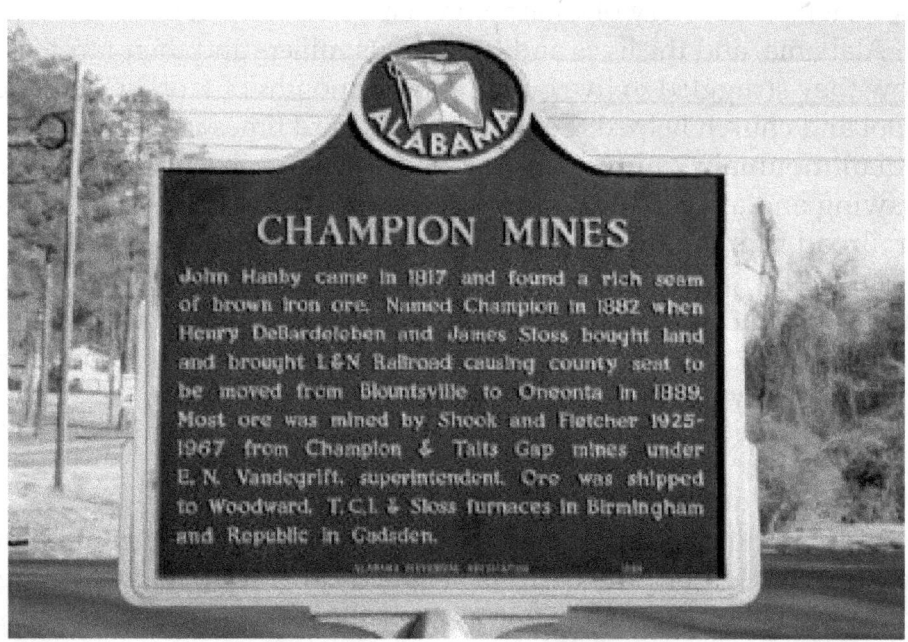

Inscription

John Hanby came in 1817 and found a rich seam of brown iron ore. Named Champion in 1882 when Henry DeBardeleben and James Sloss bought land and brought L&N Railroad causing county seat to be moved from Blountsville to Oneonta in 1889. Most ore was mined by Shook and Fletcher 1925-1967 from Champion & Taits Gap mines under E. N. Vandegrift, superintendent. Ore was shipped to Woodward, T.C.I. & Sloss furnaces in Birmingham and Republic in Gadsden.

Erected. 1989 by Alabama Historical Association.

Location. 33° 56.337 N, 86° 27.529 W. Marker is in Oneonta, Alabama, in Blount County. Marker is at the intersection of 6th Street South (U.S. 231) and Champion Road, on the right on 6th Street South.

CHAPTER ONE

MY FATHER'S MINES
By Emma Vandegrift Linder

Note - Emma was the daughter of Edgar Newton Vandegrift, superintendent of the Champion and Tait's Gap mines.

"A *goodly land*, a red chief called it, and a goodly land it is. Millions of years ago the Alabama country was begun in the great laboratory of nature." This was the opening statement of A. B. Moore's book, <u>Alabama and Her People,</u> written in 1927. It was a slow and costly process by which this Alabama country was made but it was worth it. It was brought into the modern era of "geologic history" with soil, climate, flora, mineral wealth, and esthetic beauty that made it an El Dorado of opportunity and charm.[1] Water powers were also plentiful; there is a magnificent river system. Perhaps nowhere in so small an area can there be found such a variety and abundance of mineral wealth as in Alabama. These were certain to attract the white men of the western world who were pioneering for homes.

The county of Blount in this good land can be described in the same way. Older than the state of Alabama, Blount County covered such a large area it had to be divided with the neighboring counties of Jefferson, Cullman, Walker, and Marshall. Created February 7, 1818, Blount[P1] was named for the Tennessee [P2][P3][P4]governor who had sent militia under Andrew Jackson to punish Creeks for the Fort Mims Massacre. Jackson fought and won the Creek War and many of Jackson's men came back to Blount as the first settlers. As many as fifty-one soldiers of the American Revolution finished their lives with their descendants in Blount County and have been given Memorial Markers here.

Gabriel and John Hanby were among the men of the War of 1812 under Andrew Jackson whose descendants became identified in more or less degree with the coal, iron, and railroad makings of the state and building up the Birmingham District. John settled in the hill country near the locality known today as Blount Springs and Mt. Pinson, where he put up blacksmith shops. Hanby was by birth a Virginian, born in Henry County in 1774, and was brought up to follow the machinist's trade.

Early in the 19th century he emigrated, as so many Virginians did, to Fayetteville, Tennessee, and on Jackson's call for men in 1812, enlisted. Stories of General Jackson's campaigns contain accounts of blacksmiths in his army and the making of crude implements and shoes for the horses in the cavalry forces, made from Alabama iron. John Hamby, following his trade, made what the young country needed most at that time: knives, rifles, guns, and pistols; and he brought

---

[1] *History of Alabama and Her people, A.S. Moore, Volume 1 - The American Historical Society, Inc., 1927, page 1*

Champion Mines

his sons to the business. He prospected around considerably and found a pocket of brown iron ore in the neighborhood of what is now Oneonta.[2]

John Hanby's brother, Gabriel Hanby, did not continue his occupation of ironworker but turned early to trade and politics. He built his three-story log home on a picturesque site near the Locust Fork of the Warrior River and began his contributions to Blount County and Alabama. Soon after his arrival he was easily elected a delegate to the Constitutional Convention which met in Huntsville on July 5, 1819.[3]

Concerning iron ore, George Powell wrote in his History of Blount County in 1855:

> *In Murphree's Valley there are fine beds of iron ore on vacant land within four miles of good water power, also a number of good mill seats, while white limestone, good firestone and good coal bed one foot thick are within one-half mile of the ore beds. Yet with all these natural advantages for making iron, Blount pays annually for thirty thousand pounds of Tennessee bar iron.*

With all these natural advantages of our county, her soil, coal, iron, and limestone, her climate and mineral springs, her prosperity can never increase in a high degree until she is rendered more accessible. Powell continues, prophetically:

> *If[P5] the future prospects of Blount, for advancement depend on this. But the inevitable laws of trade and commerce must, in a short time open a highway to and from all these natural resources. The people of Blount should, therefore, remain on her soil, and not suffer a mania for emigration, to lead them off. When the Atlantic and Gulf Coasts open a market for the products of our soil and mineral wealth and when railroads have made our mountains accessible to the seekers of health, we will see the lands of Blount, not rated at three and five dollars per acre but at twenty-five and fifty. We will then see capital-seeking investment among us.*[4]

The State's great mineral wealth was scarcely touched before 1860. Calhoun and Autauga counties led in the manufacture of iron, with Calhoun having a blast furnace in 1836 and the great industrial county of Jefferson was not accredited by the federal census report with a single iron industry. The people of means in the towns and the planters supplied themselves largely from northern and European factories.[5]

---

[2] *Story of Coal and Iron in Alabama, Ethel Armes 1910 page 23 University Press Cambridge, U.S.A.*
[3] *A Profile of Gabriel Hanby 1786-1826 - Blount County Historical Society, page 5*
[4] *Blount County Glimpses From the Past - Blount County Historical Society, page 20*
[5] *History of Alabama, A.S. Moore 1934, page 288 - Benson Printing Co. Nashville, Tenn.*

Champion Mines

During the years that intervened between the American Revolution and the Civil War, the energies of the Southern states were devoted more to agriculture, especially King Cotton, and the sizable iron-malting resources of the Deep South were largely unexploited. This was said to be a root cause in the defeat of the Confederacy; not until the Civil War was half over did the South create the factories to furnish its troops with much needed guns and ammunition, and to replace parts and maintain railroads to move troops. Thus, the South lost the war due largely to lack of support from King Iron, whose absence was felt in daily small nuisances at home as well as on the battle lines.

Disastrous as it seemed to Southerners at the time, defeat in the Civil War drove home a hard lesson that revitalized this whole section of our country in a short time. The people soon began to exploit the immense iron and steel-making resources in the mountains of Tennessee and Alabama. Among shiploads of new world settlers were some experienced English iron workers. During the early eighties the coal and iron industries boomed in the environs of Birmingham. The rich iron ore find of John Hanby was acquired by Major Tom Peters and Colonel James Sloss; Peters later sold his half-interest to Henry F. DeBardeleben who had induced a number of prominent ironmasters to come down from Kentucky and Tennessee. The mine was then named Champion.[6]

A general Railroad Incorporation Act of Alabama, December 29, 1868, gave the railroad companies free and liberal privileges to run their lines over anybody's land. Another Act on February 21, 1870, agreed to furnish aid and credit of the state of Alabama for the purpose of expediting the construction of railroads in the state.

Champion Mine was worked under contract for several years by J.W. Worthington and Company. One branch of the L & N Railroad Mineral Line went no farther than Champion ore mine. The train went there every night and back to Oneonta to spend the night, leaving for Birmingham the next morning.[7]

Between 1886 and 1890 there was a "boom" in the mineral belt. Along the railroads where the mineral resources seemed superior, towns were laid off with streets and avenues. Oneonta did not develop from a cross-roads store. It was designed and laid off as a city. It was the outgrowth of the development of some of the mineral deposits in the vicinity and the building of the Birmingham Mineral Branch of the L & N Railroad. The Birmingham Mineral Railroad, the L & N Railroad, the minerals, soils, and stones gave birth to the city of Oneonta.

William Newbold, L & N Superintendent, with headquarters in Birmingham, enjoyed Oneonta, "the place of many rocks" and named the new city for his home of Oneonta, New York.[8]

It had been discovered that this particular spot was surrounded by the same minerals that have made Birmingham an industrial center: red iron ore on the west, brown iron ore on the east, limestone on the west and north, coal on the

---

[6] *Story of Coal and Iron in Alabama, Ethel Armes 1910, page 23*
[7] *The Heritage of Blount County - Blount County Historical Society, page 235*
[8] *The Heritage of Blount County - Blount County Historical Society, page 234*

north, south, east, and west with shale and sand in close proximity. The presence of these natural resources attracted the attention of such men as John W. Worthington, DeBardeleben, and other industrialists and the town sprang into existence and developed rapidly for a time. Real estate brought fancy prices. Then the depression of the nineties struck the country and the development stopped for a time.

This history was recorded by Dr. M. C. Denton in The Southern Democrat and published in the Heritage of Blount County: During this time the town was incorporated under the charter of 1891, the county seat had been moved from Blountsville to Oneonta in 1889, and a new courthouse and jail had been built in 1890. Great quantities of brown iron ore were being shipped from Champion Mines two miles east of Oneonta. Red iron ore was being mined at Compton in the southwestern part of the county, coal was being shipped from Inland, known at the time as Swanson, eight miles south of Oneonta, where now is constructed the Inland Dam to furnish Birmingham water supply. Since that time coal mines have been opened, in addition to the iron ore mines, near Champion and Tait's Gap, a few miles east of Oneonta. From the beginning Oneonta has been fortunate in having many excellent business men come to our city and establish stores and other businesses. Oneonta has, since the beginning, been a city of home owners and home lovers.

In 1964 the iron and steel industry was described in a "Know Your America Program" as our mightiest industry and it was stated that, "Of all the iron mined within the United States since Columbus discovered America, one-third has been extracted from our soil during and since World War II." According to an estimate in 1959, the steel industry in the United States was processing each year nearly 110 million tons of ore, of which about 73 million were from United States sources and the balance imported. Imports increased until mines could not economically compete and by the seventies mines were phased out; in 1970 Sloss Furnace discontinued processing pig iron[P6].

It was in the period from 1920 to the 1960s that the Champion Mine was at full production with night and day shifts much of the time, operated by Shook and Fletcher Supply Company with Ed Vandegrift as Superintendent. Vandegrift came from the red ore mines of Red Mountain in Birmingham to the brown iron ore mines at Tait's Gap, north of Champion.

Vandegrift came from the red ore mines of Red Mountain in Birmingham to the brown iron ore mines at Taits Gap, north of Champion. Deep veins of ore extended from the foot of Straight Mountain at Taits Gap to Champion which had been mined at intervals for many years.

An abundance of water from mountain springs was impounded in lakes for washing the strip-mine ore and separating the iron ore from the soil and rocks. Washers were built at Tait's Gap and Champion. It was these years of improvements in machinery and methods which permitted the vast shipments of ore to Sloss and other furnaces.

Employment was high. Experienced workers came from iron ore mines in Talladega County to fill many positions. Champion and Tait's Gap were thriving communities with many family members working for good income for the time. Young men were employed at an early age as "mud ball picker "along the conveyer belts at the washers or as water boys and advanced as they learned skilled duties. Few families remain in this area without some member who once had employment at Shook and Fletcher Mines.

In 1926 the need for electricity to operate the washers, pumps and other machinery brought the Alabama Power Company to Champion and Tait's Gap Mines, electrifying[P7] the communities in the area. Stem, Delco and Homelite plants had sufficed until that time. The advantages of electricity were demonstrated in the Vandegrift home when the mining company supplied the latest in electrical appliances for home use: a large electric range, a Server electric refrigerator (when others knew only the Frigidaire), a sink with an electric dishwasher built in, and later a Carrier Food Freezer.

Alabama Power Company was a pioneer in many areas. As in the laws governing right of way for early railroads, property owners gave the same rights for power lines and were happy to do so. The power company had been first to establish a New Industries Division in 1921.[9]

This division encouraged new industry to come to Alabama by telling outsiders about the state's abundant resources and dependable electricity network. The original Sloss furnaces of 1881 were dismantled and replaced by those now standing on the site and electric power was introduced there in 1927 to 1931, thus mechanizing the iron-making process; Sloss continued to produce pig iron until 1970.[10]

Technological changes were also[P8] taking place in mining and washing ore. Mule-drawn scoops for moving ore had long since given way to steam shovels to dig and load ore onto cars pulled by Dinkey railroad engines. Later, trucks called Euclid's were used for hauling the tons of earth. Large cuts were made in the hills by draglines which stripped the earth as they moved on monstrous mechanical feet.

As mined land was abandoned it was acquired by Vandegrift who developed it into pasture and cropland on Cherry Hill Farm where an outstanding herd of Polled Hereford cattle was produced. Lakes left from the ore washing process are numerous. Gravel and fines[P9] from the separation of the ore from the muck are still being shipped to cement plants and some gravel is used for building county roads and in local construction of building foundations. Mining land along US Highway 231 incorporated into the city of Oneonta gave rise to real estate development into excellent home sites and an area for industry and business. A park was developed for use in connection with the Blount County Agriculture and Business Center to provide facilities for livestock and overnight camping for

---

[9] *The History of Alabama's Energy, Alabama Power Company, page 2*
[10] *Sloss Furnaces Museum Brochures*

herdsmen[P10].

With the phasing out of mines due to imports, the death of Vandegrift in 1987, and the closing of Sloss Furnace in 1970, John Hanby's Champion Mine closed another chapter.

So . . . Blount County progressed from a wilderness frontier inhabited by Indians to a prosperous and rapidly developing diversified economy. Agriculture and Industry[P11] met here in Alabama's vital valley.

Champion Mines

CHAPTER TWO

OUR FATHERS' MINES
By Aulden Woodard

I was one of the miners' kids. I was Champion's paperboy and my daily route took me from one end of the camp to the other. The L&N evening passenger train kicked off my paper bundles of The Birmingham Post in front of the old vacant Commissary Store at 4 p.m. on weekdays. Sometimes the miners would be waiting for me in order to get the news, ask about local news, and carry messages for them. I was seven years old, earning what I could for my Mom.

Historical Champion

This story is about Champion, a mining camp two miles northeast of Oneonta, and its miners and their families. Alongside US Highway 231, on the northeast side of Oneonta, Alabama, is an Alabama Historical Association State Marker. If you stop and read it, it tells a little of what happened here.
It reads:

> *John Hanby came in 1817 and found a rich pockets of brown iron ore and named Champion in 1882, when Henry DeBardeleben and James Sloss bought the land. James Sloss bought the land and bought the Louisville and Nashville (L&N) Railroad, causing the county seat to move from Blountsville to Oneonta in 1889. Most ore was mined by Shook & Fletcher in 1925 to 1967 from Champion and Tait's Gap mines under Ed Vandegrift, Superintendent. Ore was shipped to Woodward, T.C.I., and Sloss Furnaces in Birmingham and to Republic in Gadsden.*

However, that doesn't tell about the hard working miners and their families, the hard times, and the good times in the lives of the miners in this mining camp.

John Hanby was a Tennessee soldier who marched south with General Andrew Jackson to battle the Indians. As the troops crossed Red Mountain, Hanby found iron ore on top of the ground and in a creek bed. He remarked, *When we come back through here on the way home, I think I'll stay here and mine this red ore*, speaking to another solider. Hanby did return and obtained[P12] land for future iron production.

The Southern Democrat paper reported the Champion news on April 2, 1908:

> *Champion is on a boon[P13]. The pump houses will soon be completed to pump water for the mining process. The railroad has been extended nearly round the hill to the coal chute, and the grade will be completed in about two weeks.*
>
> *The washer will be ready for washing ore in about thirty days. The steam*

shovel will soon be in operation. There are between sixty and seventy hands at work now.

Girls, there are some of the loneliest boys here you ever saw. If you want a sweetheart, come to Champion. There are also a lot of girls here.

Boss Shell, of Chepultepec, visited J. H. Whited and family Sunday.

Oscar Whited, of Hood's Cross Roads, visited Champion Sunday.

Monroe Murphree [my grandfather] is working the county roads this week.

Mrs. Rebecca Tolbert and daughter, and Mrs. Ruth Whited, visited relatives at Chepultepac last week.

Reverend Linnie Whited visited Harp Armstrong and family last week.

Champion has the best Superintendent of any works in the country. He is Mr. John Donehoo, of Oneonta. Mr. S. J. Watts is the time keeper, and Mr. Lawrence is the Commissary Store clerk. W. H. Hannah is the engineer on the steam shovel, and James Armstrong is the fireman. Tom Jotos is the engineer on the switch engine, and Tom Mashburn is the track foreman. Taking everything into consideration, Champion is wide awake and afraid[P14] to run.

Boys, you ought to be around sometimes when the big yellow wasps get after the boys. Oh! My Lord! How they make them jump.
News was reported by (signed) the Champion Shyster.

The Southern Democrat on February 18, 1909, prints headline, "Tennessee Coal and Iron Railroad Company Opens at Champion." The company store was managed by Arthur Brice.

The Southern Democrat on June 2, 1916, reported the headline, "Champion School Day" - *the pupils of the Champion School will receive their cards of promotion, June 23rd, at 9 AM. A field meet at 3 PM, will have baseball and volleyball games at 5:30 PM. Ladies of Champion will play their husbands. Supper will be at 6:20 PM and refreshments served at the 8 PM closing.*

The Southern Democrat on August 31, 1916, prints headline, "RED MOUNTAIN ORE BEING MINED. The story read:

The Oneonta Ore & Mining Co. is the name of a new corporation recently organized for the development of the Red Mountain iron ore fields near Oneonta. This company began opening up these mines several weeks ago and are now shipping about one hundred tons daily and giving employment to about 50 men and ten or twelve teams. We understand that this force is to be doubled within the near future. New openings are being made in several places along the mountain side. The ore is now being hauled in wagons but a railroad siding will probably be built in the near future. W. H. Hannah is the president of the company.

Champion lies in a small valley on the southern side of Murphree[P15] Valley, bounded by the north side of Straight Mountain (elevation 1100-1200 feet) on the southern end of the Appalachian Mountain chain, and bounded on the northern side by Red Mountain (elevation 800-1100 feet with peaks to 1313 feet) where a fire tower was built. Over Red Mountain lies Sand Valley and beyond it is Sand Mountain. Straight Mountain spreads southerly with rolling hills and flat plateaus with rich farming soil.

Spout Spring Gap

The Champion mining camp community starts from the north passage to Straight Mountain called Spout Spring Gap, named for its spring water falling down from the mountain rocks, created when Highway US-231 was cut in 1938 through the gap between two peaks which were 1100-1200 feet tall. The spring water runs from a crevice about forty feet from the maintain top. It was prepared with a small concrete reservoir at the source with a two-inch water pipe that trailed down the jagged solid granite rocks to the side of the highway.

Travelers would stop by the spring and get a drink of the cold freestone water. Many would bring jugs and barrels to fill and carry the water home with them, some from as far away as the next state. Some people thought it would cure all of their ailments and swore by it, "...make you healthy and wise." The spring was dependable even through the driest of times.

A Mr. James Harp later bought the land where the Champion Road ends on US-231 and built a store there. He diverted the water pipe about three hundred feet down to the store, so when people stopped by for a drink, maybe they would buy something from the store. Something happened to the spring; it stopped running water after so many years. Very few people stopped by after the diversion was made. An old-timer, Cedric, told me, "The water wasn't free anymore."

Champion Mines

The Company Camp Houses Built With Vertical Boards.
Personal Photo - 1939   Aulden Woodard, 8, Returns from Hunting with Rabbit in Hand

You can see the mine's[P18] red garage and storage building in the background as well as the big oak tree with our community swing. The house is replaced now, but is on the corner of Vine Street and Champion Road.

Before the highway was built, using prison convicts, this mountain gap passage had a narrow winding road that branched northeast to Mt. Carmel Church and the other branched north-northeast toward the Antioch and Robin Hill Church communities. Champion is bounded on the southwest by the city of Oneonta. Tait's Gap is located about one and three-fourths miles north of the center of Champion (the Commissary Store).

The Champion's Mining Washer Number One was built on the side of Red Mountain on a 300-400-foot ridge about three-fourths of a mile from the camp's center, along the L&N Railroad tracks. It opened using hard labor means, first picking up ore on top of the ground, then with picks and shovels and horse-drawn scoop implements.

When steam power was available, Champion got into mining in a big way. Using steam shovels, workmen would strip the overburden (rocks, clay, chert, etc.) away down to the ore pocket, hauling it with Dinkeys, small steam engines on a narrow-gage railroad track, to the washer, cleaning it with water, then shipping it to market via the L&N Railroad. My father, Grady Woodard, was foreman.

Champion Miners
First Row: Joe Beasley, Buckie Getts,
           Wess Tidwell
Second Row: Claude Payne, Jack Clements, Buck Gallagher,
           Harrison Arnold.

Jack Fendley Photo   Champion miners with "Dinkey" engine

Champion had two washers in the late twenties until hard times came in the Great Depression. The Number Two Washer, built at the lower end, was closed and the Number One Washer work fell to two days a week. Work was limited to about seven months a year because of winter rain, ice and snows.

In 1925, when electrical power was available, Ed Vandegrift, an expert mining superintendent, was asked to reopen Champion, converting it from steam power to electric.

In the early 1930's, Champion mines employed many miners who lived in the area and in the camps. Champion mines provided about $30,000 to the local economy, a lot of money to the miners in the Great Depression years.

The community was built by T.C.I.[P19] (Tennessee Coal, Iron & Rail) , having houses built with board and batten, vertical twelve-inch wide wood planks, and joints covered with three-inch strips, an unusual and very resourceful saving on

material. Drinking water was first provided by a well dug in the center of the camp. When the well water began to get low, a water tank was built over a spring on the side of Straight Mountain further down in the camp, with a one-half-inch water distribution line. It always had a weak water stream.

Company electrical distribution power lines were installed to the miner's houses with only two fifteen-Amp fuse boxes. Most houses had four drop lights. There were no meters on the water or electricity. Each miner had an outhouse. Most miners paid $9-$10 a month rent. Wages averaged between 35-45 cents an hour. Groceries at the Commissary store were high priced, and only a few could purchase food there. Some miners went to Oneonta for groceries.

The company used water stored in ponds to wash the ore. When the company used up the water or drained a storage pond, fish was distributed to the camp miners on a pickup truck. Each would take only a small share.

The community got together and bought a community mule, housed with miner Walt Arnold, at the end of the camp. The mule was used in the gardens and patches for the miners to raise food. More and more miners went to Oneonta for groceries. The Commissary Store closed in 1935 and the operator, John Sellers, opened a grocery store in Oneonta. The Commissary Store was torn down in 1937 and it's lumber reused by the mines. Two large boarding houses in the middle of the camp also helped in the 1920s, as commuting to work any distance was a problem since the miners had no transportation.

The Company opened another mine at Woodstock in 1941 and many miners transferred to that work site. In 1941, when the easy ore was mined out at Champion, it closed. The county could not provide teachers anymore so the kids were taken to Oneonta by bus. The Champion mine was later reopened with a two-man high-tech machine washer. It was closed again in 1968 after all the economically recoverable reserves were exhausted.

There have been many miners whose sons also worked at the Champion Mines. The last of those second generation sons was Verlon Tipton. Verlon, who worked in the last high-tech machine washer in Champion was the son of John Tipton and was one of the last of a long line of miners. Verlon offered a lot of information to the Champion story.

I was one of those kids who lived in Champion. I was born in a miner's shack above the L&N Railroad tracks, on the side of Straight Mountain at Tait's Gap on July 20, 1931. My father, Grady C. Woodard, a company foreman, was transferred to Champion later that year.

BLOUNTY COUNTY SOIL MAP

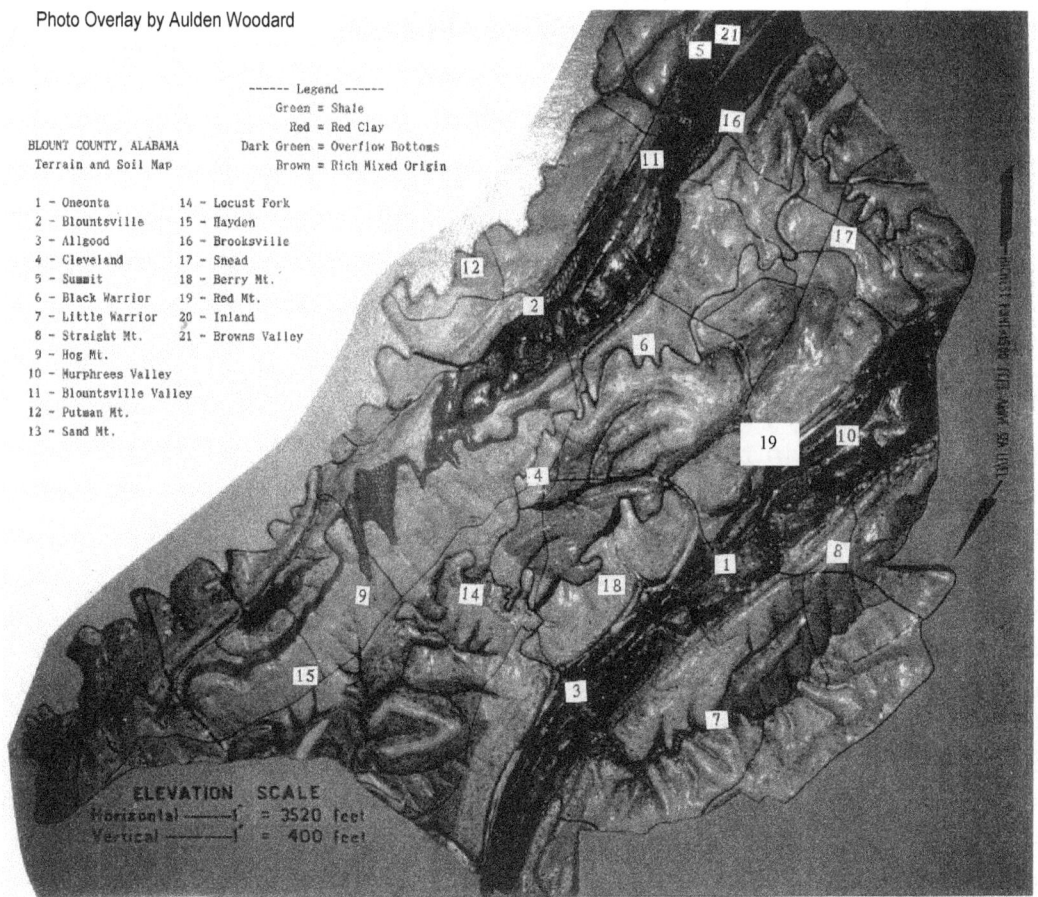

US Department of Agriculture Conservation District

The deposits of red clay show how Blount County may have an abundance of iron ore. The narrow red is shown on Red Mountain near Oneonta where the iron ore is found.

A geological survey was made in Alabama and published, entitled <u>Iron Making in Alabama</u> in 1912 by William Battle Phillips. It reports very good hard and soft iron ore in Blount County. On Red Mountain, the formation had the thickness of 225 to 275 feet and covers an area of about ten square miles. The ore has been mined at the old Compton Mines on the Northwest side of the West Red Mountain, in the West _ of Section 27 and the East _ of Section 28, Tract 14, Range 1E. At this locality, the ore was from 2-_ to 5 feet thick. The dip near the outcrop was from 30 to 35 degrees northwest, but flattened towards the Southeast. Analyses of the soft and hard ores from the Compton Mines were made by J. L. Beeson.

West Red Mountain, nearly opposite of Chepultepec, is about 300 feet high. The strata carrying the ore have, according to A. M. Gibson, the following Section on West Red Mountain, near Chepultepec, S.W. _ of N.E. _ Section 10, T.13, Range 1E.

The Compton Red Ore Mine in southern Blount County started in 1883 but closed in 1894.

CHAMPION MINERS HONOR ROLL

James Armstrong
Walt Arnold
Harrison Arnold

Bennett Battles
Cliff Battles
Willy Battles
Joe Beasley
O.L. Bellenger
Wes Blalock
Noah Breasseal

Luther Butler

J.P. Bynum

Dewy Clemmons
Jess Clements

Will Daily
Alfred Davenport
Bob Davis
John Donahue

James Engle

Hershel Fendley
Oscar Fulenwider

Webb Galbreath
Buck Gallager
Wayne Gargus
Bucky Getts
Henry Getts
Garvin Green

Solomon Gulledge
Howard Gunter
Van Gunter
Wiley Gurley

W.H. Hannah
Mr. Hardiman
McKinley Hathcock
Nathan Hathcock
Will Hathcock
Alton Henderson
Guy Higgins
Walt Hitt
Elam Hobbs
Connie Hudson
"Peg" Huggins
Ervin Hullett
Perry Hullett
Terry Hullett

Tom Jotos

Charlie Logan
Ed Lumpkin
Gilford Lumpkin
Sam Lybrand
Quinton Lybrand

Wes Mashburn
Tom Mashburn
Sires McCollem
Lavie Mitchell
J.B. Morrison

Howard Nix

Buck Payne
Claude Payne
John Payne
Julian "Crip" Payne
George Pucket

John Reneau
Mr. Rigby
Howard Roberts

Arthur Tidwell
Charlie Tidwell
James Tidwell
Wes Tidwell
John Tipton
Verlon Tipton
Joe Tuck
Edgar N. Vandegrift
Euell Vice

Mr. Waldrop
S.J. Watts
J.H. Whited
Walt Williams
Grady Woodard
Claude Young

TOTAL = 80

CHAPTER THREE

SUMMER TIME
By Aulden Woodard

Swinging in the community's swing on an old oak tree, going to school during the week, and back to church on Sunday in the same building, was about the only thing to do. We could pick plums or fly a homemade kite on the hillside, watch the trains, cars, or the wagons go by, to and from Oneonta.

It would be common to see on Saturday afternoons a wagon mule team on the road going home to Straight Mountain from Oneonta, with no one with the team. Somehow, teams would get loose in town and go home without the owner. The teams would walk slowly and move over on the road when traffic approached or pass them. Having made the trip so many times, the mules just did it themselves. It was awesome to watch them.

We could slip off to a water-filled mining hole (called the cuts) and go swimming but this would carry a penalty with it. We always got a whipping for this, because it was dangerous and we couldn't get away with it. Moms could always see the red mud on our clothes and our feet.

One day while swimming at one of the cuts near miner Walt Arnold's, one of the kids got shot in the abdomen with a .22 caliber rifle while struggling with it over taking turns shooting at a turtle. Wayne Tuck recovered, but still has the slug in him at last account. Serious restrictions were ordered. We did not go there for a long time.

Champion Church Leaders – 1936:(L-R) Willy Gurley, Grady Woodard, Arthur Tidwell, Rev. B.F. Dykes, Walt Williams and Charlie Logan

On Saturday or Sunday afternoons, the miners would play baseball at the ball field. They had helped Euell Vice and his next door neighbor, Guy Huggins, build it in the lower part of camp. Many miners were very good players like Bob Davis, the locomotive operator, and Walt Williams. The ball park was jammed with folks as they yelled for their team. Baseball was very serious in Champion and provided a great past time.

In the summer of 1936, a very sad thing happened at Champion. A rabid raccoon bit one of the camp's dogs. Buck Payne shot the raccoon and took its head to Oneonta in a shovel hanging out the car window. A meeting of the miners took place under the big oak trees in Solomon Gull Edge's yard in the center of camp. The miners decided that they would go home and assemble the dogs in camp that were possibly exposed to the raccoon. In Oneonta, the Health Department said the raccoon had rabies, but it would take a week for a test.

The miners assembled the dogs under the big oaks and led them down the tracks toward the mine's washer. A large pine tree on the north side of the track was selected. They dug a large hole under the tree and placed the dogs in it. In a bit, the shots rang out and they covered the dogs up with dirt. The camp lost fifteen dogs. The families were sad and the kids were crying.

We always enjoyed summer time after our hard work in the fields or gardens was done. There were plenty of work with our cows, pigs, chickens, and gardens. My Dad was a beekeeper and would go after bee hives in the woods. On one such trip, miners on the washer road saw a tree beehive. On Saturday, they joined my Dad to get the honey. When my Dad started his smoker, one miner, Webb Galbreath, didn't want any protection. "Bees never sting me," he said. When they returned to my Dad's house, Webb had been strung many times. Miner Will Daily saved the day by using a tobacco paste on the stings.

School always came too early. Winter was hard for all the miners and their families. We looked forward to next summer.

Champion Mines

LAYOUT OF HOUSES AT CHAMPION MINE – 1930s

LOC	NAME
1	WASHER 31 COMPLEX
2	TERRY HULLETT
3	COMMISSARY STORE – JOHN SELLERS
4	COMPANY STORAGE – GRADY WOODARD
5	COMMUNITY OAK TREE SWING
6	GRADY WOODARD
7	SAM LYBRAND
8	LAVIE N. MITCHELL
9	O.L. BELLENGER
10	SPOUT SPRINGS WATER
11	VACANT
12	NATHAN HATHCOCK
13	SCHOOL/CHURCH-MISS GILLILAND (1-5) PRINCIPAL HANTON BOWMAN (6-8) AND REV. BOB GETTS
14	SCHOOL PLAYGROUND
15	JACK CLEMENTS
16	CHARLIE LOGAN
17	MR. RIGSBY
18	ERVIN HULLETT
19	OAK TREE - ANNUAL COMMUNITY SHOT CLINIC
20	SOLOMAN GULLEDGE
21	J.B. MORRISON
22	COMMUNITY WELL (COVERED)
23	JOHN TIPTON
24	MR. WASHBURN
25	CLAUDE YOUNG
26	WILEY GURLEY
27	BUCK PAYNE
28	VERBON MORTON
29	ED LUMPKIN
30	WILL HATHCOCK
31	VACANT
32	JOHN PAYNE
33	EUELL VICE
34	GUY HUGGINS
35	VACANT MR. WALDROP
36	WALT WILLIAMS
37	CHARLIE TIDWELL
38	BOB DAVIS
39	CONNIE HUDSON
40	McKINLEY HATHCOCK
41	WILLEY & CLIFF BATTLES
42	COMMUNITY WATER TANK
43	WALT HITT
44	MR. HARDIMAN
45	ELAM HOBBS
46	WEBB GALBREATH
47	VICE & HUGGINS BALL FIELD
48	NOAH BREASSEAL
49	WILL DAILEY
50	HENRY GETTS/JAMES ENGLE
51	AUTHOR TIDWELL
52	WALT ARNOLD
53	JOE TUCK

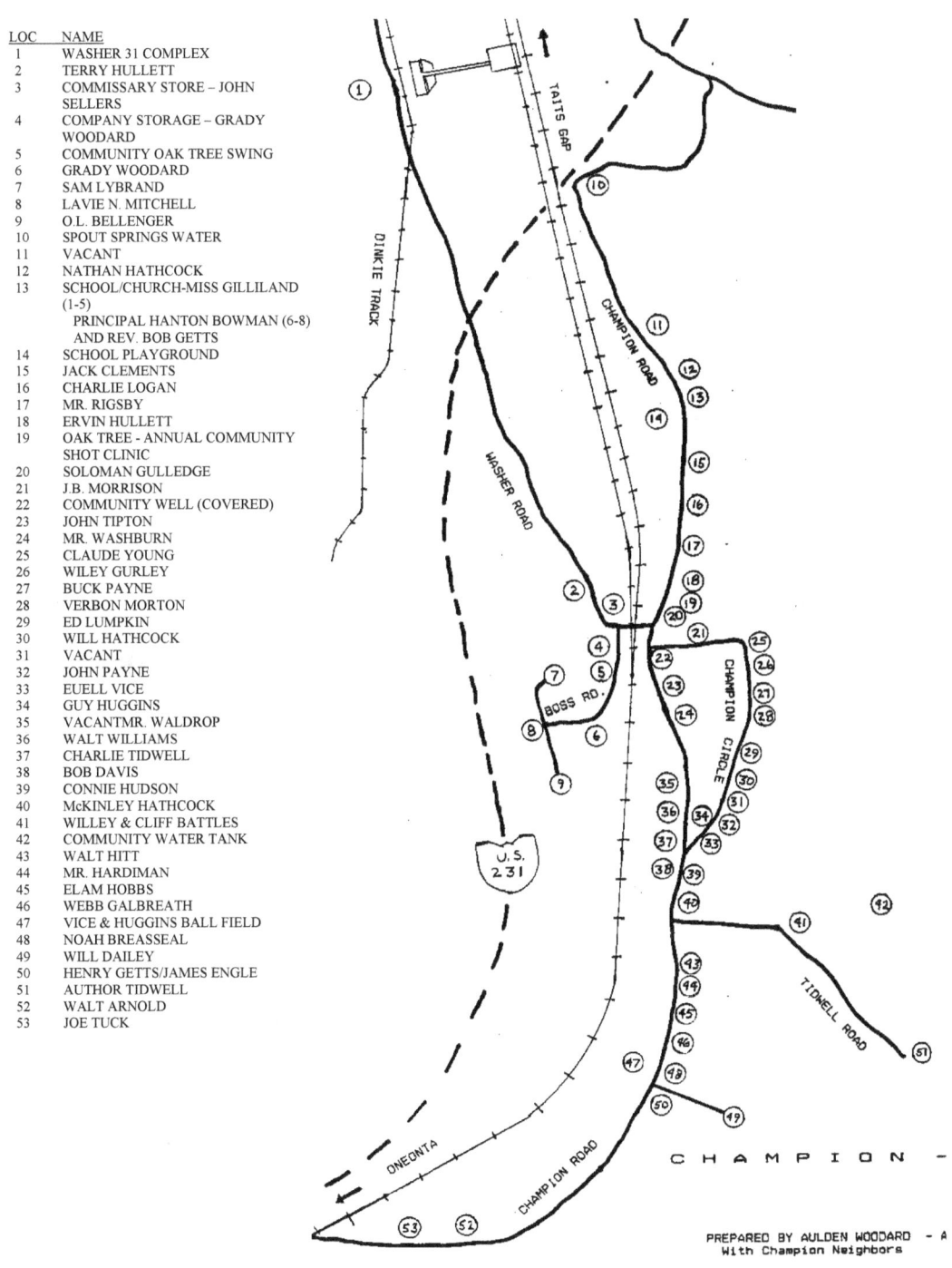

PREPARED BY AULDEN WOODARD - A
With Champion Neighbors

Map 1

Champion Mines

CHAPTER FOUR

CHAMP OF CHAMPION
By Aulden Woodard

Cora Murphree Woodard, my Mom, was one of those Tait's Gap homemakers that moved a mile-and-one-half down the tracks with her miner to Champion. She helped to organize the Champion homemakers into a formidable force to help feed their families. It was the only way to survive. She was a member of the Home Demonstration Club, a State Agriculture Department Extension Service. She was recognized as one of the best.

Personal Photo   Cora Woodard, "Champ of Champion" -

The Birmingham Post on November 16, 1937, printed headlines, "ONEONTA WOMAN CANS BIG VARIETY OF FOODS." The story read:

Mrs. G. C. Woodard, Oneonta, Route 4, has done a splendid job in food preservation this year. She has canned 783 jars of food.

The canned food includes the following products: okra, beans, peas, squash, turnip greens, kraut, tomatoes, corn, soup mixture, tomato juice, grape juice, peaches, pears, apples, plums, grapes, blackberries, huckleberries, apple butter, pear and pineapple butter, peach preserves, pears, cucumber, green tomato, and watermelon pickles, also tomato catsup, chow-chow, chili sauce and meats.

Mrs. Woodard is the only one in the little mining camp where she lives that owns a pressure cooker and she is so proud of it that she doesn't like to lend it, so she cans for her neighbors. She has canned 372 quarts for her neighbors and friends. They have paid her in vegetables and this accounts for her being able to can 783 jars for her family. After the first winter rains, Mr. Woodard, who is a miner, will be dismissed from work until the spring season. Therefore, he and his wife decided to can enough food for them this winter when they have no income.

The garden space of Mrs. Woodard's home is not very large, but she had it fertilized last Fall and had it turned and has canned about 400 jars of vegetables, fed her family, sold $2.25 worth of greens, and has given away about six bushels of beans, one bushel of squash, one bushel of cucumbers and 50 bunches of turnip greens.

She has in her garden now, turnip greens, beans, peas, okra, collards, cabbages, mustard, rutabaga, onions, and lettuce. Mrs. Woodard said, "I don't know how we existed before I learned through the extension service how to live-at-home," the paper stated.

Later, my Mom had a Saturday 30-minute radio show in Birmingham on canning food. She had about 4,000 cans of food in a large room in our house at Champion. Late one Saturday after dark, we returned from a trip to Oneonta. We found a large truck at the back door and the lights on in the house.

The driver alerted, the truck sped away with all of her cans of food. Someone must have listened to the radio show I guess. It was a terrible thing for us as we had very little to eat that winter.

Four men lost a leg and another one was crippled at the mines. My father lost a leg in 1928. He made it back to work in the spring of 1932. The company did give him his house rent free, the rest was up to my Mom during the Depression years. With a disabled husband, a crippled daughter, a two-year-old son and a newborn (me), she worked hard and got the job done.

Personal Photo. Cora Woodard's Pantry – 1937

Chapter Five

A KID'S STORY
By Aulden Woodard

In 1936, I was five years old and always sat on the railroad bank in front of our red house at the spur track switch to the mine. I became friends with the black switchman, George, on the freight train that picked up the ore cars in the evening while the passenger train went by.

In the mornings, when George's train went up the track's long grade in front of our house, George would jump off the slow train, and I met him with a half-gallon of cold buttermilk for the engine crew. On the evening return, George met me at the switch as he let the train back up into the ore mine spur track. George would give me back the milk jar for the next day's trip, with twenty cents in it.

George and I had short periods for talks and I would ask him about what was up and down the tracks, about the outside world. He gave me a lot to think about as he described how it was when a person got out on his own to work in the hard times.

George told me stories about the outside world, and warned me about bad people taking little kids away. He caught me one evening with my Mom's big butcher knife trying to whittle on a stick of stove wood. He asked me what I was making. I told him I was making a toy train engine like his so I could remember him by. He was very concerned about me cutting myself. He taught me how to use a knife and to do whittling. Next day, George brought me a nice train engine.

I asked George many times if he would take me with him out of Champion. He always said, "Your parents love you and you need to stay home and help them. Besides we need your buttermilk every day." I loved old George a lot.

One evening, the train ran off the end of the track and wrecked in front of me as I sat on the bank by the switch. I looked up and saw the men in the engine cap trying to climb out of the windows. The engine would tilt to the right then to the left and then to center. I thought the big steam locomotive was going to fall over on me. I felt someone grabbing me under the arms and wildly throwing me up in the yard. It was old George, who had to throw me to get me out of the way. George was trembling as he asked if I was all right.

The Railroad thought a five-year-old had turned the switch and asked my dad to pay for the accident. It turned out that old George had forgotten to throw the switch while talking to me and they fired him right away. I felt very bad about that and then the Railroad told me I couldn't sit by the track anymore. I had lost a dear friend.

Two months later, one Sunday evening when the Commissary Store was closed for the Sabbath, the evening train was due southbound about 4:30 p.m. I was swinging in the big oak tree swing and noticed no one was around anywhere. I started to thinking.

In the past, to stop the train, a customer would take some rolled up newspaper or a white flag, stand on the tracks in front of the Commissary Store and flag down the approaching train by waving the flag. The train engineer would give short blasts on his engine horn to acknowledge the flag and start slowing for the stop. I had seen people do this and run back to the store's porch and place the white flag (or newspaper) under the edge of the floor of the porch.

The train was coming and I could hear its horn blasting for the crossing warnings at the store. I decided it was payback time for old George, my fired friend. I quickly ran and retrieved the newspaper rolled flag and stood on the tracks and starting waving the white flag. The train engineer gave some short blasts on his engine horn and acknowledged the flag and started slowing for the stop.

I yelled, "This one's for you, George." I quickly ran and stored the flag under the porch and went around down the side of the store and to the back door of the old abandoned Commissary store to hide.

The train came to a complete stop, the conductor stepped down and placed his little wooden booster steps down on the ground and looked around. He looked and looked and could not find a customer at all. He looked at his watch and waved a signal to the engineer to proceed down the tracks. I could hear the conductor saying some very bad words, standing in the passenger car door, hollering and spitting as the train went roared off.

I was afraid he could see or hear me rolling on the ground with joy on the bank behind the corner of the store. The trainmen knew me well by sight. But I got away with it and the train was late for Oneonta, the next stop. I said, "George, wherever you were, that one was for you."

A school/church was built at the northern end of camp. Later in the late 1920's, an additional room was added. Grades 1-8 were taught at the Champion School by two teachers. Teachers were provided by the Blount County Education Department.

I went to the school for the first and second grades. Groups of kids walked up the road each morning and were joined by others along the way. Most arrived at the school at the same time. There were a lot of discussions along the way and also during the walk home each evening. Sometimes, we would pick plums or blackberries alone the way home.

I ran away from school a lot. The mines would shut down and the miners looked for me everywhere. Sometimes, I would be up a tall tree watching them walk under me. I had an accident at the community swing at the age of four with a bad head injury. Alone at the swing, I managed to get into it and started pumping. It went sideways and soon I hit the tree head on. Mrs. Lavie Mitchell found me as she was driving up Red Hill to her house. Sleep walking and nightmares became common. I finally got better in the second grade. What a relive that was for my family and the miners.

Personal Photo. Champion Kids at the Church-School – 1936. My sister, Jerry, standing in the door, had the only Brownie Camera in Champion; it was used for these pictures.

Saturday Movie Memories

I remember the task of going to Oneonta to see the Saturday western movies when we had money. For a kid living in Champion, going to the Strand Theater for a ten-cent admission ticket and five-cent popcorn was great! Walking the two miles to Oneonta down Champion Road, along the railroad tracks became dangerous so we would walk the dirt road by the tracks to High School Street.

In 1937-1939, construction on the new US 231 Highway had started at Jonah's Restaurant on Ala. Highway 75 and over Champion's Red Hill (now called Harvey Hill) for a "short route to Florida." They were cutting through Faith Stricklin's cow pasture (now Charlie B's). In Champion, they started blasting through the Sprout Springs Cut and worked across the mine's road from the Commissary Store.

Convict labor from the Guntersville Prison Camp was used to construct the new highway. We saw hundreds of convicts and chain gangs working alongside heavy equipment. We were fearful of the convicts and equipment while walking along Champion Road on our way to the movies. The chain gangs were singing and making grunt noises while working. Guards on horseback with shotguns were everywhere.

We spent most of the day watching the movies and, when we got bold enough, we walked back to Champion. These trips gave us a chance to escape the misery from a one-teacher elementary school and the rigors of Appalachian life in a mining camp – now just our memories!

Halloween 1941 Brings Heartache as School Burns

Our Champion School closed in 1941 and we were bused to the Oneonta

Grammar School, something very new for us. The new school had running water, indoor restrooms, playgrounds and a lot of teachers.

October 1941 had many great memories and one sad memory of the new school.

The historic public Oneonta Grammar School opened on Monday, Sept. 11, 1911. It was across the street from the old Sawyer Sawmill (Jonah's Restaurant and Witt Motel) on the corner of AL 75 and U.S. 231, on Second Avenue. The sawmill was noisy!

Sixty-eight years ago this Halloween (2009), on Friday, October 31, 1941, horror and grief overwhelmed us! A fire burned down the Oneonta Elementary School; it still haunts us when we remember it on Halloween. Our school principal was Ernest Bynum; our 12 teachers were Geraldine Carter, Mattie Cornelius, Madie Ellis, Wilma Eller, Mary Lou Howard, Ruby Johnson, Hattie Pass, Ruth Powell, Lura Rickles, Etta Stone Moss, Mattie Daily, and Ellie Turner. The kids from Champion School had just begun attending the school.

That Friday, we assembled in the auditorium and Mr. Bynum greeted us with announcements and a program for Halloween. There were songs and music and a stage full of ghosts and goblins. The school's Halloween party was at 7 p.m. and everybody was planning to be there.

That night at the Halloween party, we crowded in to fish for prizes and candy apples and played many games. But it was the little space under the stage floor, at the rear with an outside door, where the "Haunted House" was. One was blindfolded and led through a dark, creepy, spooky place where you bumped into ghosts and your hands were put on cold eyeballs and brains; this drew many screams.

Overnight a fire started in the cafeteria under the sixth grade and burned the school all the way around to the second grade. Fire trucks came from all over to help put out the fire. Sparks and coals of the fire rose thick and fast and were carried by the wind over the greater part of Oneonta's residential section. Never had there been a worse fire since the downtown fire on May 13, 1915, when six stores burned on First Avenue from the Citizens Bank corner to the last store at the block's end.

The following Monday, the First Baptist Church took in the third, fourth, and fifth grades and the Lester Memorial Methodist Church took in the first, second, and sixth grades. The churches used their facilities for a temporary school. It took two years to rebuild the school. There was no money for the auditorium and desks were replaced with tables. We moved back into the new school in 1943. The Bird School kids were bused to our school that year. The disruption caused many kids to do poorly and many failed their grades.

Our Champion School Teachers

Year	Teacher	Grades	Start	Cert	Pay
1930	Principal Marvin Burnett	6-8	7/21/30	E	$525.00
	Lola Bowman	1-5	7/21/30	E	$430.00
1931	Principal Hampton Bowman	6-8	7/20/31	E	$472.50
	Lola McCoullough	1-5	7/20/31	E	$409.50
1932	Principal Hampton Bowman	6-8	10/24/32	E	$240.00
	Lola McCoullough	1-5	10/24/32	E	$216.00
1933	Principal Hampton Bowman	6-8	10/23/33	E	$390.00
	Lydia Campbell	1-5	10/23/33	E	$333.00
1934	Principal Clarence Huie	6-8	10/15/34	C	$437.00
	Mrs. Lola McCullough	1-5	10/15/34	D	$441.00
1935	Principal Clarence Huie	6-8	10/14/35	C	$406.25
	Iona Greer	1-5	10/14/35	D	$390.00
1936	Principal Hampton Bowman	6-8	10/14/36	C	$455.00
	Ruth Morris	1-5	10/14/36	C	$325.00
1937	Principal Hampton Bowman	6-8	10/18/37	C	$560.00
	Roxie Belew	1-5	10/18/37	D	$525.00
1938	Principal Hampton Bowman	6-8	10/15/38	C	$560.00
	Mrs. Bonnie Ruth Walden	1-5	10/15/38	D	$490.00
1939	Principal Bonnie Walden	6-8	10/16/39	D	$560.00
	Evelyn Gilliland	1-5	10/16/39	D	$441.00
1940	Principal Bonnie Walden	5-8	10/17/40	D	$596.00
	Evelyn Gilliland	1-5	10/17/40	D	$441.00
1941	School Closed – Kids bused to Oneonta – Bus Driver Curtis Fox 1-3/4 months @$20.00 plus gas. 7 months - $341.00				

Our Oneonta Teachers

Oneonta School Teachers	Years Taught
Principal Ernest Bynum	1940-43
Principal Alfred L. Baines	1944-45
Cecil Warren	1945
Geraldine Carter	1940-41
Mattie Cornelius	1940-45
Madie Ellis	1940-42
Wilma Eller	1940-45
Dura Gilliland	1940
Mary L. Howard	1940-44
Evelyn Hunter	1940
Ruby Johnson	1940-43
Haddie Pass	1940-41
Ruth Powell	1940-43
Lura Rickles	1940-45
Etta Stone Moss	1940-45
Mattie Lee Daily	1941-42
Ellie Turner	1941
Mrs. Ernest Bynum	1942-43
Lorraine Carter	1942-43
Katheen Graves	1942
Meta Bains	1943
Lula Belle Wilson	1945
Rachel Chandler	1943
Nancy Arnold Clark	1943-45
Pauline Fendley	1943
Essie Fendley	1944
Edita Murphree	1944
Vernamae McPherson	1944
Ida Whited	1944-45
Alma Dingler	1944-45
Ona Moody	1944
Vianna Garner	1945
Rosalea Thrasher	1945
Margaret Bynum	1943-45
Otto Woodard-Bus Driver	1941
Teachers Each Year: 1940 – 13 1941 – 14 1942 – 13 1943 – 15 1944 – 16 1945 - 14	
Champion school kids arrived in Oneonta in 1941. Bird school kids arrived in Oneonta in 1943.	

Champion Mines

CHAPTER SIX

THE BIRMINGHAM MINERAL RAILROAD
By Aulden Woodard

Blount County had iron ore to mine and the pig iron furnaces in Birmingham wanted the ore. The only way to get the iron ore to the furnaces was to build a railroad to the mines in Blount County, as had been done around Birmingham's Red Mountain. The Birmingham Mineral Railroad operated in an oval loop around the big city.

Drawing by Lois Stephenson – Pinson,

The Birmingham Mineral Railroad, owned and run by the Louisville & Nashville (L&N) Railroad, and the South & North Alabama Railroad, was later taken over by the L & N, virtually monopolized the iron and coal traffic in the Birmingham area.

Starting in 1889, with the completion of the rail line from Boyles Railroad Yard near Birmingham to the Champion Iron Ore Mines, the L&N Railroad, also known as the Birmingham Mineral Railroad, began railroad freight and passenger service to Oneonta. By 1905 the railroad through Oneonta ran to Attalla and Gadsden to the northeast, connecting with the Alabama Mineral Division of the Louisville & Nashville Railroad.

Originally the L&N provided passenger and freight service to Oneonta and the L&N depot provided the passenger ticketing and waiting area. In addition, the railroad's agent had space in the depot to handle freight billing and railroad-related operational work. Railroad passenger service ended in 1951.

The growth of railroads to haul minerals developed a record number of startups and endings: the Louisville & Nashville Railroad, 1871-1983; L&N Birmingham Mineral Railroad, 1884-1904; L&N Oneonta and Attalla Railroad, 1900-1905; Cheney Railroad, 1989-1996.

It was May 28, 1905, before the 26-mile link of the Alabama Mineral Railroad was completed between Attalla and Champion, providing a more direct route to Birmingham via the L & N's Alabama Mineral Railroad.

The Birmingham Mineral Railroad was built to Chepultepec (now Allgood) to provide Birmingham with limestone, lumber, cotton, pulpwood, coal, and produce; it was known as the "Breadbasket of Birmingham." The railroad extending from Mattawanna's spur track to Cheney's Limestone Quarry was also completed from Chepultepec.

The L&N Railroad depot was located in downtown Oneonta behind the Garner Hotel. In 1977 the depot, known to have been standing in 1889, was moved from the downtown area to its present location at the Oneonta Recreational Park. A piece of history captured, it is maintained by the city and used by widely varying groups.

L & N Depot after it was moved to the Oneonta Recreational Park in 1977

From 1920 to the 1960s, the Champion Mine was at full production with night and day shifts much of the time. It was operated by Shook and Fletcher Supply Company with Ed Vandegrift as superintendent.

Cheney Lime & Cement Company, founded in 1903, built a spur track from the L&N Railroad. A vertical kiln for quicklime production was built at the site (Greystone), and General Offices were established adjacent to the plant at Chepultepec, Alabama (later renamed Allgood).

Railroads were much interested in coal, of course, as all locomotives were steam-powered, and wood-burning models had been found to be unsatisfactory. The L&N shrewdly guaranteed not only its own fuel sources but a steady revenue stream by pushing its lines into the difficult but coal-rich terrain of eastern Kentucky and also well into northern Alabama. There, the then small town of Birmingham had recently been founded amidst undeveloped deposits of coal, iron ore, and limestone, the basic ingredients of steel production.

The arrival of L&N transport and investment capital helped create a great industrial city and the South's first postwar urban success story. In the first half of the 20th century, the railroad's ready access to very high-grade coal eventually enabled it to boast the nation's longest non-stop run, nearly 500 miles (800 km) from Louisville to Montgomery, Alabama without refueling.

By 1912, the Louisville and Nashville Railroad (L&N) had four passenger trains running between Birmingham and Anniston (through Oneonta) with sixteen whistle stops. Two of the daily trains, Numbers 45 and 47, ran south, leaving at 6:10 a.m. and at 2:20 p.m. The two north bound trains were Numbers 45 and 44, leaving at 12:50 p.m. and 8:25 p.m. This did not include the freight trains on the tracks.

People were very busy with their travels because it was exciting to ride at fast speeds, see the country, and then tell about it. Sometime later the passenger trains were reduced to one train, making one round trip daily.

In The Blount Countian, on October 20, 2010, headlines read, "Historical marker to be unveiled Saturday."

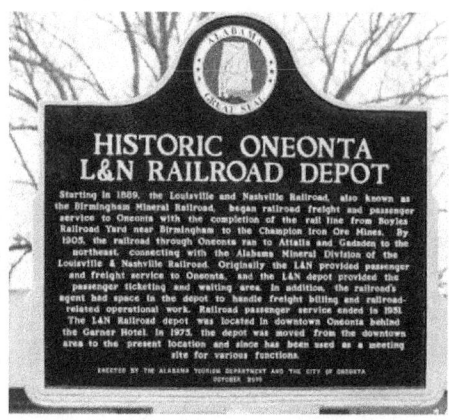

Following is the wording on the historical marker:

> Starting in 1889, the Louisville and Nashville Railroad, also known as the Birmingham Mineral Railroad, began railroad freight and passenger service to Oneonta with the completion of the rail line from Boyles Railroad Yard near Birmingham to the Champion iron ore mines. By 1905 the railroad through Oneonta ran to Attalla and Gadsden to the northeast, connecting with the Alabama Mineral Division of the Louisville & Nashville Railroad.
>
> Originally the L&N provided passenger and freight service to Oneonta, and the L&N depot provided the passenger ticketing and waiting area. In addition, the railroad's agent had space in the depot to handle freight billing and railroad-related operational work. Railroad passenger service ended in 1951. The L&N railroad depot was located in downtown Oneonta behind the Garner Hotel. In 1975 the depot was moved from the downtown area to the present location and since has been used as a meeting site for various functions.

The origin and development of Oneonta was due to the coming of the

Birmingham Mineral Railroad, a part of the Louisville & Nashville Railroad. Because of the presence of iron ore, limestone, and coal in the area, there was always a great potential for development. The early settlers recognized the value of the natural resources and the transportation provided by the railroad made the potential to become an iron and steel industry a supporting reality. As the settlers built their early homes, Oneonta's neighborhoods took shape around downtown and around the L&N Railroad depot.

There are many historic sites in Oneonta, including the Depot, the Garner Hotel building, and the historic Tin Town area. Oneonta proudly celebrates the connections to history and small towns and downtowns. The L&N depot at Oneonta Recreation Park, a piece of history captured, is maintained by the city and used by widely varying groups.

The Birmingham area was home to the nation's third largest concentration of ore mines in the United States. Iron makers discovered that Red Mountain ore, when "sweetened" with brown ore, produced superb foundry iron. Brown ore was needed and the Birmingham Mineral Railroad was extended north into Blount County to the Champion and Tait's Gap Mines to get the source.

The coming of the railroad soon proved to be the most significant event of the era, as well as moving the county seat to the new town of Oneonta. Blount County was known as the mineral region of Alabama. Oneonta was the county seat of Blount County. It was first opened to settlers from eastern Tennessee, and the region took shape between 1830 and 1900. The Birmingham Mineral Railroad started in 1888.

Transportation was afforded through the medium of the great Louisville and Nashville railroad, which traversed the land from the north to the south. This railroad penetrated the heart of the famous Murphree's Valley, and along its route, from one terminal point to the other, were vast deposits of coal and iron. The road proved to be of incalculable advantage to the population residing in the region through which it passed. Such was the attractiveness of this region that it quickly allured a population as soon as its resources of mine and soil were known.

In Blount County and the adjoining counties which lay along the railroad, the value of the lands diminished the further they were from the line of communication. Land could be purchased in the county at prices ranging from $5 to $35 per acre at the time.

There were 34,320 acres of government land in Blount County and the combination of the Champion and Tait's Gap mines and the railroad brought a big boom to the area. They played the most important part in Blount County history. The L & N Railroad gave the city of Oneonta the Railroad Station Depot. The Depot was known to have been standing in 1889.

Alan Cheney, Sr. purchased the railroad in 1989, after abandonment of the line, and tried to make it a viable short-line rail operation with which to continue his Lime and Cement Quarry shipment needs, and because it meant so much to the county and its people for ninety-nine years.

After her husband's death in 1995, Mrs. Cheney closed her husband's Lime and

Cement Quarry Company. When Cheney abandoned the Cheney Railroad in 1996, by pulling up all rails and ties, they gave up any rights of to the 57-mile-long corridor. The 110-foot-wide corridor reverted back to the original land owners.

Personal Photo. Abandoned L&N Railroad, the end of an era.
Looking from High School Street - 1997

Champion - Railroad Era Ends

After 99 years, the railroad closed because the Champion Mines and the Limestone Quarry had closed. A huge economic impact to Blount County, the value lost cannot be calculated. From my old Blount County High School Street in Oneonta (picture above), that looks down the track at the old Railroad Depot Station location. To the right, behind this picture, my High School Senior Building once stood. It was there that the railroad cross-ties and rails were stored and loaded onto large trucks on their way to salvage. A motel now stands where the main school building once stood.

It was an awesome sight to see and I was there when the last Cheney little red engine made its final ceremonial run. The little engine moved slowly with its loud whistle blowing; its drivers waving to the crowds gathered along the railroad, villages, and towns to say good-by. Flashbacks of their own memories of Champion and the railroad came to many. I rode the train north from Champion and south to Birmingham on the last car platform many times; the view of folks living by the railroad as I did, I will not forget!

The New Champion Village Today

To go back to Champion today and drive down Champion Road, one can see a new life and a community of residents and industry.

Champion Road and its two streets, Ladie Lane and Vine Street, have thirty-one residents, six companies, and one church. Some of its people do know about the old village there and its history. It is so awesome to see that the old abandoned community lives again!

CHAPTER SEVEN

THE SUPERINTENDENT
By Aulden Woodard

Edgar Newton Vandegrift,
Miner and Mine Superintendent
1885-1967

Edgar "Ed" Newton Vandegrift married Aurelia Mae Alverson on November 11, 1910, in Coal City, St. Clair County, Alabama. Vandegrift was an iron ore miner and superintendent in Tait's Gap and Champion mines in Alabama.

In 1916 Vandegrift began mining red iron ore with Thomas Worthington on the mountain behind what is now Park Avenue in Oneonta, Alabama. By 1918, he was mining red iron near Birmingham and on July 26, 1920, he arrived at Tait's Gap when Worthington began mining brown iron ore, completing his last move in mining iron in Alabama.

Vandegrift came from the red iron ore mines of Red Mountain in Birmingham to the brown iron ore mines at Tait's Gap, north of Champion. Deep pockets of

Champion Mines

ore extended from the foot of Straight Mountain at Tait's Gap to Champion which had been mined at intervals for many years. An abundance of water from mountain springs was impounded into lakes for washing the strip-mine ore and separating the iron ore from the refuse. Washers were built at Tait's Gap and Champion. Improvements in machinery and methods through these years permitted the vast shipments of ore to Sloss, T.C.I, Republic, Woodward, and US Pipe.

From 1920 to the 1960s the Champion Mine was at full production with night and day shifts much of the time. It was operated by Shook and Flectcher Supply Company with Ed Vandegrift as Superintendent.

Employment was high. Experienced workers from iron ore mines in Talladega County came to fill many positions. Champion and Tait's Gap were thriving communities with many members of families working for a good income for the time. Young men were employed at an early age as "mud ball pickers" along the conveyer belts at the washers or as water boys; they advanced as they learned skilled duties. Few families remain in this area without some member who once had employment at Shook and Fletcher Mines.

The Southern Democrat December 13, 1928:

VALUABLE INDUSTRY

While there are many industries shut down in the Birmingham district, and others mining half-time and one-third time, there is one in the Oneonta district that has been running day and night for more than a year in order to fill the demand for its products.

We refer to the Shook & Fletcher Supply Co. operations at Champion and Tait's Gap. This concern, under the management of Mr. Ed Vandegrift, has expanded from a small beginning a few years ago until now it employs about three hundred men and has a monthly payroll of $30,000.00.

From these mines comes some of the highest grade brown ore to be found in any part of the country. It is a grade of ore that there is always a market for. When the mines were first opened in 1888, the ore produced was regarded by mining engineers and chemists as the best ore of its kind in the world. From this the place derived its name Champion--the champion ore of the world.

After being in operation for a number of years, for some cause work was suspended for a time. When electric power became available through the Alabama Power Co., the mines were re-opened by Mr. Vandegrift and both the Champion and Tait's Gap giants were electrically equipped.

This industry means much to Oneonta and the surrounding country. It's annual payroll equals one-fourth the entire cotton crop of Blount County.

While Mr. Vandegrift does not employ many high-priced men they are good honest men who meet their obligations and men who are never in the courts for violations of law. Another thing that makes this a valuable industry for Blount County is the fact that nothing used in the development or obligation of these mines has been such that could be bought outside of the county that could be bought in the county

An industry like this, though small, has a distinct effect on the county. We need more of them and we need more men like Mr. Vandegrift.

Vandegrift's method in operating the mines during the Great Depression was his own. When he had no ore orders, he helped feed his miners. He kept paying his salary, used the money to buy seed and used the mine lands for share crop farming with his miners. He was found out and the owner cut his pay off "since he became a farmer instead of miner." Vandegrift may have felt omnipotent to violation of law. His miners thought he was a Hero and "He walked on water," many said.

CHAPTER EIGHT

FATHER AND GRANDFATHER
GRADY WOODARD AND MONROE MURPHREE, MINERS
By Aulden Woodard

Grady Woodard, Foreman

A miner from 1924 to 1941, **Grady C. Woodard** was born on January 14, 1902, and died on December 2, 1985. He married Cora Murphree (descendant of Daniel Murphree) on September 24, 1922; both were from Straight Mountain farms. Woodard first worked at the Tait's Gap Mine in 1924 and transferred to Champion Mine in 1931 as Washer Foreman. He was assigned the Shook & Fletcher/T.C.I. Office (house) in Champion after the Champion co-owners moved out. It was painted red and had a red garage-storage building across the Mine Road from the Commissary Store at the foot of Red Hill. It was where the bookkeeper, O. L. Bellenger and foremen Lewis Gunter and, later, Lavi Mitchell had lived. He had 35-40 men working on the Washer. Woodard earned $.45 cents an hour, five cents more than his men.

Accident!

At the Tait's Gap mine, Grady met with a terrible mining accident in March, 1928, which cost him his left leg (below the knee) and over two years of lost work. It was Grady's job, when a railroad car was filled with iron ore, to release the air brakes and let the car roll downgrade and to spot it on the spur track. He would then go upgrade and roll the next empty car and spot it for loading.

A two-week new hire named Mr. Jones from Texas had been asking to roll the ore cars down; he was told he didn't have the training because sometimes the air brakes leaked down and the mechanical brakes were slow and hard to work. Mr. Jones decided to mount the second filled ore car and let it roll. Grady was standing on the car connection lock with his left foot swinging in the jaws. The Washer made a lot of noise and Grady could not hear the second ore car rolling. When the two cars slammed, connecting jaws, his foot was smashed in the jaws and locked. Jones jumped down and ran away telling no one about the trapped miner hanging on the ore car. Finally, a miner went to check on Grady's progress and found him in agonizing pain hanging on the car. It takes the power of a locomotive engine to bump the cars apart. All of the miners on the Washer rushed to put long steel edging bars under the steel wheels of both cars to push them apart; they could not budge them.

Then, a southbound locomotive came down the main tracks. It was switched to the spur and bumped the ore cars apart to free Grady. After lying for three hours in the topside office, he was carried to Oneonta to a doctor. He was later transferred to a Birmingham hospital where he stayed for thirty-two weeks receiving thirty-two operations to save his foot. His leg was amputated below the knee and he was sent home to Champion. The mine's office let his family live free in the ten-dollar a month rental house during his hospital stay. The Company also told him his job would be waiting for him when he returned from sick leave. One of Woodard's men, Sam Lybrand, was assigned to work as foreman. All this happened during the Great Depression. He had three kids, including a new born and a handicapped daughter. He learned to cut hair for ten cents a cut with his injured leg on a pillow in a chair. He finally returned to work in 1931 with an artificial leg.

The neighbors helped out as much as they could and caught fish in the company's ponds for them. Grady's wife Cora planted gardens in the fields around Champion with the use of a company mule assigned to the mining camps. She then canned the food from her harvest. Miners could only work seven months before bad weather set in and winter came. Mrs. Woodard was a member of the Tait's Gap Home Demonstration Club and had the only pressure cooker in the camps. She organized the other mining wives to help in home canning the food.

To supplement their income, when <u>The Birmingham Post</u> newspaper District Manager wanted Champion deliveries, she had one of her boys do the job. That job was given to me, Aulden, at age six, delivering the papers on a bicycle. I saw the camp miners on a daily basis from 1937 to 1941 as the Champion paperboy.

Woodard was not the only man hurt on the job of rolling down the ore cars. Later, in 1938, Sam Lybrand, his acting foreman at Tait's Gap, continued at Champion to work the ore cars. Sam was holding his head down looking inside an ore car while releasing it to roll and the ore hopper's chute caught his head inside the car wall and mashed it into an oval shape. Sam survived, had a lot of surgery and time in the hospital but he could not work again.

New Housing

In 1936, when a better house became available across the tracks, on the corner of

Champion Circle (now Vine Street), Woodard moved into what had been J. B. Morrison's house after he moved to Oneonta's McCay Avenue to a rock house.

The old red Company Office was never used again as it was falling down in the rear. Woodard's new neighbors were Solomon Gulledge on Champion Road and Claude Young on Champion Circle (now Vine Street). This location was considered to be the center of Champion; it involved a community well and had a baseball field on the opposite corner.

In 1940 the Champion Mine owner Ed Vandergrift told Woodard at an often-scheduled Sunday dinner that the mines were playing out and warned him to start looking for another job. Woodard enrolled in the Alabama Power Company's electrical and refrigeration school in Oneonta. When he finished school in 1941, he left the mines, moved to Oneonta, and established a shop on Main Street. The Champion Mines closed in 1944.

Woodard was hired to do equipment maintenance by the Gordon - Patton Hospital and he was a member of the Alabama Hospital Engineers Association. It was noted that Ed Vandegrift had asked Woodard many times to help him start up the new Champion Mines since the ore had not played out after all. Woodard was reluctant to accept the job because he thought the risk of the former co-owners may not have known that information. The new startup of the Champion Mines was in 1961 and closed in 1968. After Woodard retired, he lived on Canal Street in Oneonta, Alabama.

Champion Mines

Monore B. Murphree, Miner

My grandfather **Monroe B. Murphree**, a descendant of Daniel Murphree who founded Murphree's Valley, was also a Champion miner. He worked by making roads, railroad beds, truck roads, and steam shovel paths. His equipment was a mule team-drawn road-blade scraper. When he quit mining, he purchased a road scraper, assembled a team of mules, and did private work for the public along with his farming.

Often, Murphree would do driveways and work on dirt roads over Straight Mountain and Champion, wherever donations were raised. Murphree had a farm off of what is now US Highway 231 near Antioch Church.

CHAPTER NINE

MINER VERLON TIPTON, "MR. CHAMPION"
By Krissy Taylor, Granddaughter

Foreword - The families of Tipton and Morton played an important part of the Champion Mines history with son Verlon Tipton, father John Tipton, Uncle Milton Tipton, Grandfather Arthur Morton, Uncle Verbon Morton, Uncle Ollie Morton and Uncle Lonnie Cornelius. Although miners called Verlon "Tip," everyone knew he was "Mr. Champion."

Arthur Verlon Tipton was born on August 29, 1929, to Lillie Ovada Morton-Tipton and John Wiley Tipton. He was reared on Straight Mountain in the area of Mount Carmel Church. Vernon's life began with a connection to the T.C.I. Ore Mines and later, Shook & Fletcher, when they were in Blount County the first time around at Champion. Verlon had a lot of family that worked at the mines including his grandfather Arthur Morton (Foreman). After Shook & Fletcher took over the mines his grandfather Morton continued his work and other members of the family joined him. Verlon's father John Tipton (washer) and uncles Milton Tipton (washer), Verbon Morton (bulldozer operator and shovel operator), Lonnie Cornelius (jig operator), and Ollie Morton (driller) all worked in the mines until they left Blount County for Odenville, Alabama.

Father's Accident

Verlon's life was also affected by the mines because his father John Tipton was one of three men that lost a leg working in the mines. In 1929, when he was just 22 years old, John lost his leg below the knee while loading ore into a train car from the washer, before Verlon was born. John was compensated by Shook & Fletcher Mining with a trade course of his choice in plumbing, electrical, and a barber course to teach him a trade. John elected to take the barber course. He was given a wooden leg. Despite the loss, he continued his work at the mines as a washer, for a total of seventeen years, until the mines closed and moved to Odenville, Alabama.

Family and Camp Life

Many of the miners and their families lived in the camp houses that the company had built. John and his family moved into a camp house when Verlon was about 6 years old. They lived there a little over a year; John and Lillie missed living on the mountain where they both had been reared and spent their adult lives up to that time. In the time that John and the family lived in the camp they experienced a little of what camp living was like. Verlon attended the two-room school house in the first grade; Ms. Roxie Bellow was his teacher.

The family also had a garden in the camp like many of the other families. Many of the other families had milk cows and chickens for eggs. John and his family bought their milk from one of the other families in the camp. In the camp there was also a community potato patch that the company had planted and tended; Mr. Walt Arnold planted and tended the patch. The day before it was time to plow up the potatoes, he would go through the camp and tell everyone to get the "tubs" ready for the next day and "to come and get all you want."

Mischief

The kids in the camp were always getting into different kinds of mischief. Verlon's fondest memory of playing at the camp was when he and some of the other boys, including Lawrence Mitchell, Olden Hobbes, and Herbert Hodges, would soap the Dinkey tracks and then hide in the woods and wait for the Dinkey to come through. The soap on the tracks would keep the Dinkey from climbing the smallest of grades on the track and the driver would have to get out and clean the tracks with sand paper and then rinse it off with water. Of course, this was very frustrating to the driver Mr. Charlie Tidwell. If he could have found those boys . . . let's just say it would not have been a good thing.

Mines Close and Reopen

Not long after John moved his family back to the old home place on Straight Mountain the original Champion mines closed and all operations moved to Odenville. Many of the families moved to Odenville to continue work, others went to work for the red ore mines in Birmingham, and some elected to find other employment. John decided to put the barber trade, which he had learned after he lost his leg in the mines, to good use and began his new career as a barber. Many of the other members of Verlon's family went to Birmingham to work for T.C.I. in the red ore mines; they commuted daily.

In the 1950's mining came back to Blount County with Shook & Fletcher Mining at Tait's Gap. Several members of Verlon's family came back to the mines - Ollie Morton, Lonnie Cornelius, and Verbon Morton. This is where Verlon Tipton began his work in the mines. Verlon was working for a produce company delivering produce and he wanted to try something different. The only way to get a job at the mines in those days was as the old saying goes, "someone had to retire or die."

Verlon heard from his family at the mine that Mr. Burt Deerman was going to retire and he should get his name in the hat for the job. So he first approached Mr. Vandegrift about a job in 1959, but Mr. Vandegrift told him, "you don't want to work in all this mud." Verlon expressed a sincere desire to take the job, but Mr. Vandegrift was adamant in his decision. Verlon was also adamant about getting a job at the mines, so he proceeded to talk with members of the Board of Directors at the local bank – Almus Hanson, Emery Lowry, and Andy Moses – about helping him get on at the mines; Mr. Vandegrift was also on the Board. All of them knew Verlon very well and trusted that he meant what he said. Several days later, after the next board meeting, Verlon got a phone call from Mr. Vandegrift telling him he had a job as soon as he could work out a notice at his current job.

Verlon found someone that night to take over his produce route and began work in Tait's Gap mine the following morning, February 12, 1959. When he began work at the mine he was a truck driver, driving a F8 Ford truck making about $1.75 an hour working ten hours a day six days week. The miners were paid for a forty-hour week and then for twenty hours of overtime; this is what made the jobs at the mine so appealing to so many, the overtime. After about six months of work Mr. Vandegrift came to Verlon and told him that he was surprised that he was still there and that he was doing a good job.

While working at the mines in Tait's Gap, Verlon worked under several foremen. A few of them were: Bruce Morrison, Tom Green, and Jim Dickie. Verlon spent most of his time driving a truck to transport ore and mud. But when Shook & Fletcher decided to move operations from Tait's Gap back to Champion in the early 1960's, Mr. Vandegrift asked Verlon to clear the land for operations. He was told to take one man and a chainsaw and start clearing trees; he took Hoyt Berry and went to work. They were told to cut the trees down and that someone else would come through and haul them off. Howard Dover and Herb Gargus were given that job and in the process Mr. Dover suffered a broken leg after getting hit in the leg with a log. It took about three months to get the trees down and ready to setup equipment. The "New" and final Champion opened in 1961.

After the mines opened at Champion Verlon went back to driving a truck, he drove a different kind of truck when they moved to Champion; some of the equipment was upgraded. New trucks were part of that upgrade; the new trucks were called Eucs (Euclids). These trucks were 22 tons and would carry 22 tons of ore.

One day not long after moving to Champion, in mid-December 1961, Verlon was out cutting wood at the request of Mr. Vandegrift and cut off his left big toe. Hoyt Berry came to his aide, but after Verlon took off his boot, Mr. Berry saw what had happened and fainted. Mr. Vandegrift sent Verlon to Doctor's Hospital and Doctor Hobbes reattached his toe and Doctor Ira Patton treated him for the injury. Unfortunately they were not able to do anything to repair the brand new pair of boots that he ruined.

Dam Breaks

In the mid-1960's the dam on the mud pond broke; the mud pond was where the tailings were deposited that had been washed off the ore. The dam broke early one morning before the first shift started. Soon after Verlon got a call at home telling him that the dam had broken and that part of Highway 132 was closed. Verlon went to the office to find several members of the Shook family already there with Mr. Gunter and several of the other foremen.

Everyone was shocked and distraught over the disaster and did not know what to do. Then in comes Mr. Vandegrift and tells them not to worry, to just get to work getting J.T. Tolbert's place cleared up so that he could get his delivery trucks out. He owned an egg company, not connected to the mine, and he needed to be able to deliver his eggs. So that is what they set out to do. With equipment, shovels, and man power they got his trucks out and back on the road. Several homes were moved off their foundations and were flooded with mud. Only one person, Mrs. Clarence Elrod, had to be rescued from her rooftop and, thankfully, there were no deaths or injuries due to the flood. Some believe that the disaster was minimized by the fact that when the damn broke the mud split and went in two different directions.

Mr. Vandegrift found a new way to dispose of the mud and operations continued at the mine. In the late 1960's the men were gathered and told that the mines would be ceasing operations and that if they could find jobs then they should grab them up. Verlon was able to find employment in Birmingham at Vulcan Rivet and Bolt Company and left the mines before the actual closing in 1968. When he left the mine after approximately eight years of service he was making $ 1.99 an hour working a sixty-hour week. The rest of his family who were still working at the mines - uncles Verbon Morton, Ollie Morton, and Lonnie Cornelius all retired when the mines closed.

Verlon looks back on his days at the mines, the times growing up, and working there as an adult with fond memories and friendships that have lasted a lifetime.

CHAPTER TEN

ARTHUR TIDWELL – LEADER AND SYRUP MAKER
By Gene Tidwell

Example of Tidwell Sorghum Syrup Mill

Arthur Tidwell was not an iron ore miner but a coal miner and a farmer on the edge of Champion; he was a sorghum cane syrup maker too! If Champion was a public community and not a mining camp, Arthur Tidwell would have been Mayor. His leadership in the community, its church, and his warm, caring affection for its people, was well remembered. His son Ralph Tidwell did become Mayor of Oneonta and served for many years. The Tidwell family was a large one with many sons and one daughter.

Mr. Tidwell was not only the Champion Church leader, he was also our Sunday school teacher. He booked our visiting preachers and when he couldn't get them, he gave the sermon himself. He led our choir and congregation in singing with his Great voice. His favorite song was "Give Me That Old Time Religion" that rocked the people at his sermon closure.

Tidwell raised a cane crop and harvested it each year. He then borrowed his father's sorghum syrup mill and set it up. People of Champion came after dark, with their hot biscuits, to buy sorghum syrup and help run the mill. It was set up near the well-lit back yard of miner Henry Getts' house on the corner of Champion Road and the road to Will Daily's farm. It was unofficially the **Champion Syrup Festival**.

Mr. Tidwell enjoyed people coming out and talking to him about it and tasting it. He also liked the fact it was a family project. His family helped all along the way.

The planting was from late May through early July and it took about three months to mature. The family planted a few acres of the crop every year. Once they harvested it, the tops were removed and the remainder was run through a mill to extract the juice.

The mill itself was an interesting setup. A mule-drawn wooden arm was rigged to turn the mill. Cane was fed into the rollers that squeezed the juice out which was then collected in a wooden keg with a spigot. A hose was connected to the first cooking pan over the fire; it was plugged with a corn cob and when juice was needed, it was released. The remaining cane was used to feed the family's cows, keeping the cow feed bill low.

A long fire stove was set up with walls. Opened at one end, it was vented through a smokestack on the other end. The cooking was done in a series of copper pans where the juice flowed from one pan to another, each one traveling over fire that was hotter. The art of keeping the fire just right required a helper constantly working the fire to perfection under all the cooking pans. This was hard to do in a long firebox using a poker rake. The management of the fire and adding wood to a targeted area was key. As the cooking progressed, Tidwell told the fireman which pan needed attention.

Tidwell would stir and strain the syrup throughout the heating process using a strainer to skim residue that built up at the top. The residue was tossed out. The syrup thickened along the way until Tidwell determined it was perfectly stringed. The sweet syrup smell was evident as it heats.

People watched Tidwell lift the long-handled dipper and watched the syrup drip from it, the way his father taught him to do – to string it – holding the dipper up and watching the juice run off it several inches below. When the drop stringed out, it was about time. Once the syrup hit that level, a spigot was opened and the syrup was strained through cheesecloth into a bucket collected by the fireman.

The end result was delicious syrup made through the same process Tidwell had learned from his father Manley Tidwell while growing up. The annual **Champion Syrup Festival** was always a big success!

During this time, one could hear laughter from the people that gathered, telling stories with merriment while the children played. Sometimes Cliff Battles would bring his banjo and someone else, a fiddle or a mouth harp. It was a gala affair and provided a release for the miners' problems and worries. It was the fun thing to do!

There were many stories told about the large Tidwell family, such as the one about a dry and very large, round, deep hole in their front yard, built by using a long pole walking around the sides of it to enlarge it. There was another story about "The Big Scare."

The Big Scare

One day Mae Tidwell was alone with one of her young sons when she heard a strange thumping. Being a scaredy cat with a negative imagination, she thought somebody was trying to rob her. She ran out on the porch to yell for her neighbor, but she was so afraid that her yell, "Aunt Jenny! Aunt Jenny!" came out in a weak whisper.

She ran to a neighbor, Melvin Hill, who grabbed his gun and came running. Sure enough, they heard the thumping sound. Mae hollered, "We're gonna make them devils bleed!"

At her voice, her son Little Bob, rose up from behind the kitchen table. He had fallen asleep on the bench and when he would almost fall off, he would thump against the wall. Bob still remembers Mr. Hill standing there with his gun, saying, "Well, hello, Bob". His life was spared.

The Tidwell Family was much-loved by all.

CHAPTER ELEVEN

WILEY GURLEY, STEAM SHOVEL OPERATOR
By Sarah Gurley Richardson

Wiley

Wiley Gurley, was Champion's Steam Shovel Operator. He was born June 6, 1894, and died on September, 1975. He married Ada Phillips ion 1917; his children were Wesley, Sarah, John, and Gary.

Dinkey ore car being filled by the steam shovel while Dinkey crew (Hershell Fendley and Buck Payne) wait on back of Dinkey – 1928

Gurley came from the Talladega mines to Champion. Wiley was a community leader in Champion's school and church. He loved baseball and played on the

Champion Baseball Team.

Gurley lived in one of the T.C.I. houses on the very top of Red Hill in Champion. He wanted the kids to have a Recreation Park. One day Gurley came home with some miners, bringing an old steel cable he had replaced on his steam shovel. They worked hard building a community swing in a large oak tree at the foot of his hill next to the Shook & Fletcher Red storage garage. This was Champion's only hang-out where kids could play. Many gathered there on evenings and weekends for baseball, swinging, talking, playing, and having fun picking blue berries on the hill behind the swing.

The Red Hill Road circled in front of the swing, from the Washer Road, the Commissary Store, and the railroad crossing connecting Washer Road to the Main Champion Road. The lot between the swing and the railroad was used for a baseball field. Wiley Gurley made all the kids of Champion Happy!

When the Champion mines closed in 1944, Gurley moved to Oneonta on Valley Road near David Moody's Airport. He worked at the Cheney Line Quarry after Champion. Then he worked as a Shovel Operator at Talladega, Champion, Cheney Lime and Cement, and at Birmingham Slag as a Security Guard before retirement.

Oak Tree Park

Steam Shovel

A **steam shovel** is a large steam-powered excavating machine designed for lifting and moving material such as rock and soil. It is the earliest type of power shovel or excavator. Steam shovels played a major role in public works in the 19th and early 20th century, being key to the construction of railroads and the Panama Canal. The development of simpler, cheaper diesel-powered shovels caused steam shovels to fall out of use in the 1930s.

Steam shovels weigh about 35 tons with a 1-_ cubic yard dipper. An average day's work could be from 2,400 to 3,600 tons compared to laborers' 4-_ tons a day. Crews were the operator, assistant, boiler fireman, and an oiler keeping everything greased.

Mining also benefited from steam shovels: the iron mines of Minnesota, the copper mines of Chile and Montana, the placer mines of the Klondike – all had earth-moving equipment. But it was with the burgeoning open-pit mines – first in Bingham Canyon, Utah – that shovels came into their own. The shovels systematically removed hillsides. As a result, steam shovels were used around the world from Australia to Russia to coal mines in China. Shovels were also used for construction, road and quarry work.

Photo from Jack Fendley. Wiley Gurley's Steam Shovel Crew Loading at Champion.   Hershey Fendley Crew's Dinkey Cars - 1928

Champion received a new Marion Steam Shovel in 1922. That gave them four steam shovels at two mines. The steam era ended in 1944 when Champion closed. One of the steam shovels was sold to Robbins Coal Company in 1947.

Wiley Gurley's pride was mining and baseball; he reported their games as well.

Baseball news in The Southern Democrat, August 15, 1929:

Champion lost two games to Cleveland, Saturday, August 10th, by scores of 5-1 and 4-3. Both were well played with the exception of three errors made by Champion in the first game, letting the game get away by doing so. Batteries first game; Allred and Livingston; Bullard and Allred, second game Wadsworth and Livingston; Rutherford, Blackwood and Allred. These teams will meet at Oneonta Ball Park, Saturday, August 17. Champion's team will be stronger as old John Reneau will be with us; also Luther Huey. So come out and see the contest.- Reported by Wiley Gurley

Baseball news in The Southern Democrat, August 12, 1937:

G.N. Vise, manager of the Champion ball team, having played W. T. (Dub) Hitt in every position on the field, decided Saturday to try him at pitching. As usual the young man came up with a sterling performance, beating the Gulf States Steel Co. of Gadsden by the score of 12-0. Hitt only allowed two hits during the nine innings. Batters were: Frazier, Wilburn and Robertson, Hitt and Bains. The Champion team also defeated Nixon Chapel, 10-7, Alabama City, and New Hope 9-4. They play Nixon Chapel at Champion on Thursday and Wylam on Friday of this week at Champion. You are cordially invited to attend these games. - Reported by Wiley Gurley

Champion Mines

CHAPTER TWELVE

THOMAS RUSSELL (TOM) GREEN, MINER
By The Green Family

The life of **Thomas Russell (Tom) Green** began in the year of 1896 in St. Clair County, Alabama. He was the son of Daniel M. Green and Mary E. (Mollie) Langley Green. They made their living farming in St. Clair County until his father's death at 51 years of age. Sometime after his father's death in 1907, the family moved to Oneonta and lived with Tom's oldest brother, Robert "Bob" Green and his wife.

Tom joined the U.S. Army sometime after 1914, at the start of World War I. He returned home and lived with his brother and family as well as his mother and sister Ellen until he married his wife, Nancy Jane McGurik (Magourik).

Tom started working for the ore mines sometime around 1920. His brother (Bob) was already working at the ore mines. His brother-in-law, James McGurik (Magourik) also was working at the mines in Blount County. They all lived in the township of Hoods. It is now known as Hood's Crossroads.

Tom's sister Ellen had married Jack Clements, who was also working at the same mines. Ellen and Jack also had taken Tom and Ellen's mother to live with them. Some say that Mollie, their mother, was blind from birth but others have said that she went blind after she married.

The year the mines closed is uncertain. Tom purchased a farm on what is known as Straight Mountain in Blount County, Alabama. He farmed until the mines reopened and he went back and forth to work at Tait's Gap.

He continued to work between Champion and Tait's Gap depending on which site was open. When the mines closed again Tom and others moved to Odenville to work mines there but due to lack of ore they closed. The mines reopened at Champion and Tom went back and worked there until he retired sometime in the fifties. Tom's next to the youngest son worked at the ore mines.

Tom and Nancy had eight children (six boys and two girls.) Most of the children married and continued to live in or around Oneonta. Some of the children and their families still live in Oneonta and Blount County today.

CHAPTER THIRTEEN

TAIT'S GAP MINES
By Aulden Woodard

The history of Tait's Gap started when John Hanby came in 1817 and found rich deposits of brown iron ore. It was named Champion in 1889, when Henry DeBardeleban and James Sloss brought the railroad to the mine. The ore deposits extended up the track to Tait's Gap, where a camp was built in 1920.

Most ore was mined by Shook & Fletcher in 1925 to 1967 from Champion and Tait's Gap mines under Ed Vandegrift, Superintendent.

Ore was shipped to Woodward, T.C.I. and Sloss Furnaces in Birmingham and to Republic in Gadsden. Both Tait's Gap and Champion mining camps operated at the same time.

Photo from Jack Fendley   E. D. Battles Store

Champion Mines

TAIT'S GAP MINERS HONOR ROLL – 1920s – 1950s

E. D. Battle	Herbert Gargus	Henry Parker
John Black	Tom Green	John Parker
Andrew Bird		Luther "Ludie" Parker
Lossie Bird	Oscar Hathcock	Owen Parker
Noah Breasseal	Hobert Henderson	Oslow Parker
Mr. Brown	Jessie Henderson	A. G. Poke
D. M. Brothers		Taylor Price
	Pink Jordon	Wesley Pucket
Rev. Earice Chanler		
E. H. Clements	Will King	Buck Redman
Jack Clemmons		
C. B. Cornelius	Elli Lodge	Cecil Shadie
Fred Cornelius	John Low	Harley Spradlin
Joe Cornelius	Will Lybrand	Standly Stoddard
Lonnie Cornelius		Chuck Smith
	Jessie Marshall	
	John Marshell	Jerry Tanner
Clarence Davenport	Arthur g. Madden	
D. N. (Pop) Davenport	Amos McClendon	Oscar Walker
B. A. Deerman	Clarence Moody	Walt Williams
Jim Dickie	Ray Morrison	Taylor Wittington
Pink Dicky	Ray Morrison, Jr.	Grady Woodard
Wheeler Dillard	H. Verbon Morton	
	John Moses	Edgar N. Vandegrift
Willie Franks		
	Andrew Oden	TOTAL - 53

WE HONOR THE MEMORY OF
MINER
CLARENCE MOODY, AGE 27
WHO WAS KILLED IN THE MINES
ON NOVEMBER 16, 1935
AT REST IN ANTIOCH
METHODIST CEMETERY

Tait's Gap Map – 1930s

LOC	NAME
1	Washer #1
2	Company Office
3	Company Shop
4	Joe Cornelius
5	Vacant
6	Vacant
7	Ed Vandegrift
8	Vacant
9	Vacant
10	Church
11	Ray Morrison, Jr.
12	Ernest Self
13	Ray Morrison, Sr.
14	Vacant
15	Mr. Bird
16	Mr. Bird's Son
17	Community Well
18	Arthur Parker Store
19	Vacant
20	Baseball Field
21	E.D. Battle
22	E.D. Battle Store
23	Railroad Depot
24	Vacant
25	Mr. Hartley
26	Vacant
27	Ore Loading Ramp
28	Oscar Walker
29	Vacant
30	Will Cain
31	Vacant
32	Grady Woodard
33	Vacant
34	Community Well
35	Vacant
36	Tait's Gap School
37	Vacant
38	Jessie Marshall
39	Pop Davenport

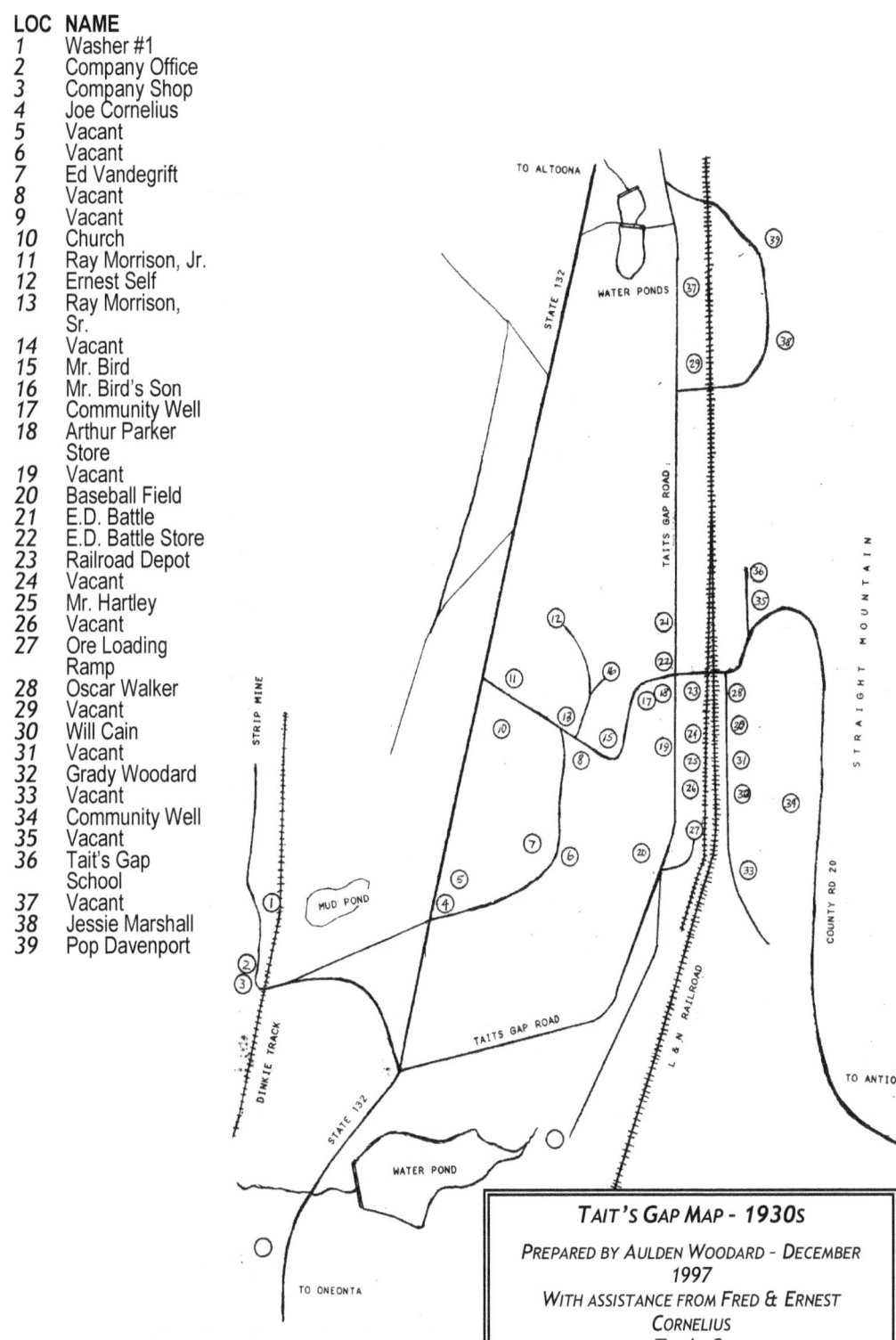

TAIT'S GAP MAP – 1930s
PREPARED BY AULDEN WOODARD – DECEMBER 1997
WITH ASSISTANCE FROM FRED & ERNEST CORNELIUS

Champion Mines

CHAPTER FOURTEEN

AURELIA MAE ALVERSON VANDEGRIFT
TAIT'S GAP MINERS' HERO
1891 - 1987
By Aulden Woodard

Aurelia Mae Alverson

Aurelia Mae Alverson married Edgar "Ed" Newton Vandegrift, on November 11, 1910, in Coal City, St. Clair County, Alabama. Ed Vandegrift was an iron ore miner and superintendent in Tait's Gap and Champion, Alabama.

Mae Vandegrift was one of those members of the Tait's Gap homemakers that lived in the mining camp near the railroad tracks with her miner Ed Vandegrift and their family.

She discussed with her husband what the wives of the miners had told her about how hard it was for the families to live. As she became close to the homemakers, she investigated ways that the ladies could improve the meals at home and make homemaking easier. Soon, Mae was holding meetings with the ladies in camp with ideas about providing more food with gardens, chickens, hogs, and having cow's milk. The group explored more ways to can and preserve food.

Mae asked her husband if the miners could go together and purchase a mule for community use, keeping the mule in the company's mule barn. With a mule, the miners could plant gardens and fields to help them during the Depression years. Ed Vandegrift agreed to house and feed their mule.

In 1935, she helped to organize the Tait's Gap and Champion homemakers into a formidable force to help feed their families. It was the only way to survive the Great Depression. Mae took guidance and information from other communities that had a formal club with the State Agriculture Department Extension Service.

Champion Mines

Mae also showed the wives of the miners how to feed their families better during the difficult off-season, winter months when the weather prevented operating the mines. In Champion, she had Cora Woodard, whose miner husband Grady had transferred as foreman from Tait's Gap, to partner with her and lead the Champion effort to improve the lives of their miners too.

Often the Vandegrift's invited the Woodard family to Sunday dinner in Tait's Gap. I knew that our fathers would talk about mining after dinner but our moms, Mae and Cora, would talk about homemaking and their Club. How could they improve and help feed the miners families? How could the use of the community mules and planting gardens and fields be maximized?

The two homemakers taught other homemakers how to can, sew, cook, and preserve food. Mae knew how to cook and sew before she joined the Club but with information from the Club, she bought her first pressure cooker and canned meat not only for her family, but also for others. Mae is what one would call a "social lady," a Hero!

Later, Mae helped to organized the formal Tait's Gap Home Demonstration Club, a State Agriculture Department Extension Service, and it was recognized as one of the best. The group met at the home of Mrs. Herbert L. Brothers under the leadership of Home Extension Agent Pauline Holland King. Mrs. Brothers was the first president. Some of the later presidents were: Mrs. Ercy Fowler, Mrs. Herbert Gargus, Mrs. Dennis Robbins, Mrs. Julian Conn, and Mrs. Jessie Marshall.

Mae Vandegrift's new enterprise developed out of United States Home Extension Service activities in Alabama. Congress passed the Smith-Lever Act in 1914, establishing the Cooperative Extension Service, whereby county agents would call on farm communities to teach farmers the latest agricultural technology and, as well, to teach home economics to women and girls. Home Extension Agents made quite an impact on the often isolated communities they served, many of which were without electrical service, telephones, or even rural mail delivery.

In the South, the boll weevil and generations of agricultural practices that left much of the land eroded and unfertile had taken their toll. The Extension Service was a means by which the federal government hoped to transmit information that would help rural people upgrade their lives.

Women as well as men were targeted in the federal government's efforts to institute rural improvement programs. Female home demonstration agents became familiar figures in rural communities. The clubs they organized provided training not only in nutrition, hygiene, and child-rearing techniques but also in craft skills useful to home-making and developing home industries. In Alabama, there were some notable successes in the teaching and production, and ultimately the marketing, of home-produced crafts.

Mrs. Woodard said, "I don't know how we existed before I learned through the extension service how to live-at-home," a quote in <u>The Birmingham Post-Herald</u>, November 16, 1937 (see Chapter 3 for more).

This story demonstrates how Aurelia Mae Alverson Vandegrift did a great job in helping both mining camps in theirs efforts to improve the lives of their miners, helping them to survive the Great Depression and the misery of a one-teacher elementary school and the rigors of Appalachian life in a mining camp.

Chapter Fifteen

TAIT'S GAP LIVES AND TIMES
By Aulden Woodard

In May of 1995, I arrived on Bird Lane in Tait's Gap at an old unpainted miner's house with buckled boards on the porch. The house and yard needed work. I knocked on the door and waited. A pleasant lady opened the door.

I interviewed Peggy, sister of miner Lossie Bird, sitting in her front porch swing in front of the Tait's Gap Baptist Church, a gentle breeze with fragrances of honeysuckles coming from the hill behind the house. She told me stories about her community.

Peggy, in her late years, told stories of the miners. The miners used to sing a song about Tait's Gap as they worked to the rhythm, tune, and tempo of "Mammy's little baby loves shortin' bread" except, the lyrics were about mining iron ore with their co-workers, for their kids, wives, and the boss man.

Tait's Gap Mine - bottom, showing train tracks and flume line waste water - 1923

Clarence Moody Killed

Peggy told of the deadly accident at the Tait's Gap Mine years ago that killed a miner, Clarence Moody, age 27, on Saturday, November 16, 1935.

Clarence married Pauline McCullough; he was a father and a family man. Clarence and Pauline worked hard to take their place in the community and were church members. Times were still hard in the

Champion Mines

aftermath of the Great Depression and the mines had received orders for more iron ore shipments and were pushing late into the winter weather break, to finish loading the last rail car.

The accident occurred next to the steam shovel at the end of the Dinkey track. A dirt bank caved in and Moody was crushed against the shovel's dipper while digging and loading ore into rail cars. It was a very sad thing to occur to a nice young man and the miners of both Tait's Gap and Champion mines were stunned. Never before had there been a fatality, although there had been disabling injuries.

Many miners paid their respects at the Emmett Funeral Home in Oneonta and the funeral at Antioch Methodist Church. A large crowd gathered at both locations and talked about raising funds to help pay the miner's funeral expenses.

The miners contributed what they could in donations and held a fundraiser at both Tait's Gap and Champion churches in what they called "Box Suppers." This involved the adult ladies preparing a box of delights such as pies, cakes or meals. The beautifully wrapped boxes would be auctioned off with the winning bidder getting to enjoy it in the company of the charming lady who made it. Some guys watched for any signs of the most attractive box preparer for clues so they could bid up on the prize box; the bidding would be wild at times. The miners were a close group and when something happens to one of them, they all pulled together to help solved the problem.

Another miner, Andrew Bird, lived next door on Bird Lane and lived the hard times all the miners had at the mines. Peggy said, "In 1930, about 300 miners worked here; by 1940 it was down to about 50 miners."

The lives and times in Tait's Gap were often reported like these newspaper articles:

The Birmingham News - April 19, 1989:

Tait's Gap a Shadow of Former Self

TAIT'S GAP – High, brown grass growing in the middle of the railway tracks rustles in the spring breeze. It is a ghost of the sounds that once made this a thriving mining community.

Tait's Gap, about five miles north of Oneonta, was in its heyday in the early part of the century when Shook and Fletcher Mining operated the iron ore mine here. At one time several hundred people lived here. Prosperity continued through World War II. Then slowly it began to change, the people moved away; today there are only about six families living here. Emma Linder, who lives north of Tait's Gap, remembers moving there in 1920. Her father, Ed Vandegrift was the mine superintendent. "There were 25 houses in the community", she said. "A school was built and children from the area attended."

Mrs. Ludie Parker remembers that there was a barber shop, a blacksmith shop, and two stores on the main street.

> *My father-in-law operated one of them. He had come here in the 1920s. That store building which dates back to pre-WWI days still stands. In the early 1930s another store opened next door, operated by the Battles family. Mrs. Ed Battles was the operator, sold not only food but also gasoline and an assortment of antiques. For years the stores prospered. The trains ran four times a day and a depot was built.*
>
> *When World War II came, some of the men who went to the military left from the depot. And some came home after it was over.*
>
> *It was in the late 1950s that things started going downhill; the iron-ore mines had closed, and late in the decade, Parker closed his store. Other things also faded out, including the school and the depot.*
>
> *Mrs. Battles kept her store open, but in the 1970s a coal truck came down the road off Straight Mountain, careened over the tracks, and smashed into the store. No one was injured and the store front was repaired. But by then, it too was fading and soon Mrs. Battles closed it. She died a few years later.*
>
> *Most of the houses built for the miners have been torn down. The school was closed. The trains stopped running. Only about a half-dozen houses remain. The Parker family lives two houses from the store. William Reynolds lives across the tracks from the stores.*
>
> *The closest stores are in Oneonta or Altoona. This is still a good place to live. But we get a lot of traffic. A lot of people use the Tait's Gap Cutoff Road here to get them to or from (US) Highway 231. We get traffic 24 hours a day.*

"One thing that has still remained through the years is the Tait's Gap Baptist Church. A tornado destroyed it in the 1940s, Mrs. Linder said. It was rebuilt with bricks and a tall steeple. It was at the end of a road, about two city blocks from the stores.

Today, on Tait's Gap's main street, the stores stand empty and silent, adorned with weathered signs - Tait's Store and Mrs. E.D. Battles Store - which herald an era gone by.

The Southern Democrat newspaper had a reporter collect the news in the mining camps and report it to the paper. This one is by the Tait's Gap reporter; it includes Straight Mountain and Champion:

Champion Mines

The Southern Democrat - March 12, 1931:

Tait's Gap News (March 2, 1931), reported by "Hoover's Step Children Dick and Dale."

Health of this community is very good at present. John Ratliff and wife are rejoicing over a new girl-number 2. Mrs. Odes Whittington is spending this week with her mother at Oneonta. The Prayer Meeting at Mrs. Fronie Mae Huie's house last Wednesday night was enjoyed by all present. Miss Dicie Hall, Miss Elnora Huie, and Earnest Huie attended the singing. At Antioch the Fourth Sunday and reported a nice time. Rev. E. D. Battles preached a good sermon Sunday night and we had a good prayer meeting also.

We have Prayer Meeting at the Church every Sunday night and have Cottage Prayer Meeting every Wednesday night. Everyone is invited that would like to come. Mrs. Willie King is recovering from the mumps. Mrs. Luther Huie and mother will leave Friday for Talladega to visit their little girl who is in the deaf and dumb school. Grady Woodard and wife were seen out riding Sunday. The school children will hate to see their teacher leave for Howard College at the end of three weeks.

Luther Nash and wife were the guests of Mr. and Mrs. Austin Whitley Sunday. Clem Hale was the guest of his sister and brother Sunday. Hermon Whittington went to Hendrix Sunday as usual. Loyd Bynum was the Sunday night guest of Miss Ilene White. Webster Galbreath and father-in-law Bob Smith have gone to Florida on a visit. Mrs. D. B., your piece, "The Unwed Mother" was certainly good.

Straight Mountain News - Folks come on with your news. It sure is interesting to us Tait's Gap folks. Mrs. Lou (Bill) Woodard of Allgood is up spending a few days with her daughter Jessie Huie and son Grady Woodard. Grady and Finis Lowe were the Sunday guests of Roy Lowe.

Champion News - We hope Champion guys don't feel hurt because they didn't get to work all last week. Tait's Gap guys' pocketbooks are kind of empty over their lost time. Ha, ha. Say, Mrs. Jacobs, you had better laugh a little over this as you are such a good laugher.

Pearl Lowe spent Sunday with Mrs. Fate Lowe. John Hartley, Fate Lowe, Woodrow Parker, Roy Lowe, and Thomas Hartley went fishing Friday night and reported a mosquito bite. Dave Davis and wife spent Sunday evening with his mother, Mrs. Jim. Davis.

Jack Clements and little four-year-old girl spent a few days with Mrs. John Lowe last week. Billie Fay Gunter was the Sunday evening guest of Imogene Tidwell. Mr. and Mrs. Elton Lowry spent Sunday evening with *his mother, Mrs. Harley Lowry.*

Champion Mines

The Tait's Gap School

The Tait's Gap School on the slope of Straight Mountain, across the railroad tracks from the two stores, played a central part of the community. The teachers worked hard to educate the area kids. Grades 1st - 8th at first, then 1st - 6th were taught by these teachers:

TAIT'S GAP SCHOOL TEACHERS
FROM BLOUNT COUNTY SCHOOL MEETING MINUTES

Name	Dates	Salary/Year
Lula Belle Wilson	1928-1929	$??
Ether Rice	1928-1929	$595.00
Cloy E. Miller	1929-1930	$480.00
Roxie Stephen	1930-1931	$525.00
Roxie Stephen	1931-1932	$472.50
Wynelle Huggins	1931-1932	$351.00
Roxie Stephen	1932-1933	$240.00
Wynelle Huggins	1932-1933	$200.00
Mattie Cornelius	1933-1934	$406.25
Mattie Cornelius	1934-1935	$455.00
Mattie Cornelius	1935-1936	$455.00
Mattie Cornelius	1936-1937	$595.00
Mattie Cornelius	1937-1938	$595.00
Audrey Cornelius	1937-1938	$490.00
Mattie Cornelius	1938-1939	$770.00

Notes: Minutes list possibly Ruby Johnson and Mamie Self.
Tait's Gap closed in 1939.
Mattie Cornelius transferred to Oneonta, 9/18/1939 at $770.00.

STUDENTS BUSED TO ANTIOCH, BIRD, AND ONEONTA SCHOOLS
1939 SCHOOL BOARD LISTS NUMBER OF TEACHERS FOR EACH GRADE

Grade	Number/Teachers per Grade
1	23
2	27
3	13
4	4
5	?
6	1
7	1
8	1
High	6
Evening	1
Colored – 1st	4
Colored – 3rd	1

CHAPTER SIXTEEN

FATHER AND FATHER-IN-LAW
WILLIE FRANKS AND HARLEY SPRADLIN – MINERS
By Margretta Smith and Willie Franks, Great-Granddaughter

l-r: Children Carnell and James with Mr. Harley J. Spradlin and Wife Glenda

Harley James Spradlin was born in 1915 to parents Tom and Lauretta Spradlin. The Spradlin family lived on Red Mountain overlooking the mining town of Altoona. At one time, Altoona was a prospering mining town that attracted families who were eager for employment. The mines were open during a turbulent time in history when the nation was experiencing the Great Depression. Though mining was hard work, families were grateful just to get a job.

In the early 1930's, Harley began working in the Tait's Gap mines. In 1932, he married Imogene Franks and continued working in the mines alongside his father in-law, Willie Franks. Both Harley and Willie would rise early in the morning, with the roosters crowing. Harleys wife, Imogene, would make breakfast and fix an extra lunch for Harley to take to work. When Harley wasn't working in the mines, he was farming.

Soon the Spradlins moved off of the mountain and settled in the Booker Branch community on the Little Warrior River. Harley continued farming and mining. However, he no longer worked in the Tait's Gap Mine; he worked in the Cleveland mines that were owned by Doctor Towns. Harley received stamps for his wages.

As time went on, Harley would eventually change his occupation and contract with the state of Alabama as a bull dozer operator for construction of public roads and commercial sites. The time he spent working in the mines contributed to his health problems. In the 1970s, he was diagnosed with Black Lung. He would later die in 1982 at the age of 66. He is buried at Wynnville Cemetery in Altoona, Alabama.

WILLIE A. FRANKS - MINER

Willie A. Franks was born in 1889. His parents were sharecroppers who moved to Blount County from Georgia. In 1911, when he was about 22 years old, he married Callie Irene Sims. Both newlyweds lived with Willie's parents in the town of Cleveland. Willie farmed with his dad for some time while in Cleveland and then moved the family to Altoona.

Around the 1930's, while living in Altoona, he worked as a miner in the Tait's Gap mines. This work was very hard but it helped pay some of the bills. While Willie worked in the mines his wife, Callie, worked as a mid-wife and an herbalist. Both he and his wife were well-known and well-liked in the community.

Not long after Willie started working in the mines, he developed rheumatoid arthritis in his hands from digging in the mines with a pick. He would eventually move his family to Oneonta and change employment. However, because of the severe arthritis that developed in his hands while he worked in the mines, his work days were limited. He later died at the age of 84.

Accidents

Tait's Gap mine accident in June, 1925 – Julian Payne, 17, had the misfortune to have his right leg lacerated by a tram car running over his leg. The flesh was torn away from his leg but the bone was not broken. He was placed in the Birmingham infirmary and hopes to return home soon.

Accident at Tait's Gap mine in March, 1928 – Foreman Grady Woodard met with a terrible accident which cost him his left leg (below the knee) and lost work. When a rail car was filled with ore, Woodard release the air brakes and let the car roll downgrade and spot it on the spur track. A new hire mounted the second ore car and let it roll. Woodard was standing on his car's connection with his left foot swinging in line with the connection jaws. When the two cars slammed, his foot was smashed in the jaws and locked. The new hire ran away. Miners could not loosen the cars. Soon a locomotive came down the main tracks and bumped the ore cars loose. Weeks later, his leg was amputated below the knee and he was sent home. After a long recovery and a new artificial leg, Woodard returned to work.

John Tipton was one of three men that lost legs working in the mines. In 1929, John lost his leg below the knee while loading ore into a train car from the washer when he was just 22 years old. John was compensated by Shook and Fletcher Mining with a trade course of his choice in plumbing, electrical, or as a barber, to teach him a trade. John elected to take the barber course and he was given a wooden leg. Despite this training, he continued his work at the mines as a washer, for a total of seventeen years until the mines closed. He moved to Odenville, Alabama.

An accident occurred at the Champion mine in June, 1938. Sam Lybrand met with a terrible accident while releasing the air brakes and let an ore-filled rail car roll downgrade and onto the spur track. Sam was holding his head down looking inside the car while it was rolling under the ore hopper's chute and his head was caught inside the car wall and mashed into an oval shape. Sam survived, had a lot of surgery and time in the hospital. He was sent home from the hospital but he could not work again.

Electric power, 1925-1944, brought more efficient washing plants and maintenance. During this phase gasoline-powered dump trucks began to supplement and replace the Dinkey locomotives.

MASTER HONOR ROLL
By Aulden Woodard

CHAMPION AND TAIT'S GAP MINERS' NAMES
REPORTED OVER THE YEARS (AS OF DECEMBER 17, 2010)
SONS, HUSBANDS, FATHERS, GRANDFATHERS AND GREAT-GRANDFATHERS, UNCLES & COUSINS

A - (8) - Clarence James Alexander, John Alexandra, Grady Alexandra, Carl Allen, Ed Armstrong, James Armstrong, Harrison Arnold, and Walt Arnold.

B - (30) - James T. Baker, Bennett Battles, Cliff Battles, E. D. Battle, Ellis F Battles, Willy Battles, L.A. Beasley, H. Beason, Joe Beasley, William Beard, Hoyt Berry, O. L. Bellenger, John Black, Denton Blalock, John W. Blalock, Posey Blakley, Noah Breasseal, Guy Brice, D.M. Brothers, E.F. Brothers, Willard Brothers, Mr. Brown, Houston P. Bruce, Ruel Bryan, Mr. Buckner, Luther Butler, Clinton G Bynum, Claude Bynum, Huley Bynum, J. P. Bynum, Jr., A. J. Byrd, Andrew L. Byrd, and Lossie Byrd.

C - (19) - James Elbert Campbell, Theorell Carlton, Louis Carroll, William Carroll, Rev. Earice Chandler, Oscar Chandler, E. H. Clements, Jess Clements, Jack Clemmons, Dewy Clemmons, Miles O. Clements, Wilburn Clevenger, Arthur Collett, J.J. Collett, Gilbert Cone, C. B. Cornelius, Ernest Cornelius, Fred Cornelius, Joe Cornelius, Lonnie Cornelius, and Joe Cotton.

D - (19) - Leonard C. Daily, Will W. Daily, Bob Davis, David Davis, James E. Davis, Shirley Davis, Alfred Davenport, Clarence Davenport, D. N.(Pop) Davenport, Bert A. Deerman, Jim W. Dickie, Lee Mancel Dickie, Phillip G. Dickie, Pink Dicky, Wheeler Dillard, Elli Lodge, Mack Dodd, John Donahue, John Dooley, Allen C. Dover, and Howard Dover.

E - (2) - Grady T. Ellis and James Engle.

F - (8) - Clifton Joseph Franks, Hershel Fendley, Sidney F. Fitts, Willie Franks, Arthur Fullenwider, Eugene F. Fullenwider, Jesse Fullenwider, and Oscar Fullenwider.

G - (20) - Ray Galbreath, Webster I. Galbreath, Will Galbreath, Buck Gallager, Herbert Gargus, Wayne Gargus, Bucky Getts, Henry Getts, Asa Green, Garvin Green, John R. Green, Nolie R Green, Okie B. Green, Tom R. Green, Howard Grissom, Solomon Gulledge, Wiley Gurley, Howard Gunter, James Louis Gunter, Roy D. Gunter, and Van Gunter.

H - (36) - Clem Hale, Dwight Hare, John Hale, W.H. Hannah, Mr. Hardiman, John F. Hartley, Lim Hartley, Thomas Daniel Hartley, Marvin Harvey, McKinley

Champion Mines

Hathcock, Nathan Hathcock, Oscar "Pete" Hathcock, Will Hathcock, Jesse Hawkins, Wilburn E. Hawkins, W. G. Hatley, Alton Henderson, Hobert Henderson, Jessie Henderson, Amos Henry, Grady Henry, Sidney W. Hickman, Guy Higgins, Elmer Hill, Eujester Hill, J. C. Hill, Jess Hitt, Walt Hitt, Elam Hobbs, Lamo Hopper, Connie Hudson, William V Hudson,"Peg" Huggins, L. G. Huie, Perry Hullett, and Terry Hullett.

J - (4) - Clarence James, Emitt Jones, Pink Jordon, and Tom Jotos.

K - (3) - John Kaufman, Scott Kay, Johnnie W. King, and Will King.

L - (18) - Grady Leak, Charlie Logan, John Low, Oscar Lowe, Edward Lowry, Elton Lowry, Jesse Lowery, John P. Lowry, William E. Lowry, John B. Louie, Ed Lumpkin, Gilford Lumpkin, James Lybrand, Ralph Lybrand, O.,M. Lybrand, Quinton Lybrand, Sam Lybrand, and William W. Lybrand.

M - (21) - Arthur G. Madden, Tom Mashburn, Wes Mashburn, Amous McClendon, Sires McCollem, Hulgier McDaniel, Bill Manning, Jesse Franklin Marshall, John C. Marshall, N. Vadion Maynor, Herman Minshew, Lavie Mitchell, Clarence Moody, Elvin L Moody, John B. Morrison, Ray Morrison, Ray Morrison, Sr., Arthur Morton, H. Verbon Morton, Ollie Morton, W.M. Morton, John Moses, and Monroe B. Murphree.

N - (1) - Howard Nix.

O - (1) - Andrew Oden.

P - (24) - Henry Parker, Ira Parker, John E. Parker, Luther "Ludie" Parker, Owen Parker, Oslow Parker, Buck Payne, Claude Payne, Clint Payne, John Payne, Julian "Crip" Payne, Clyde E. Payne, Robert Payne, Otis Perrin, Geo. Donehoo Phillips, Louis Phillips, Rufus Monroe Phillips, A. G. Poke, I. T. Price, Charles Puckett, George D. Puckett, Wesley Pucket, Clarence L. Putman, and John W. Putman.

R - (12) - Hanson Reaves, Buck Redman, William E. Reeves, John Reneau, Donald Rice, Howard Riddle, Mr. Rigby, Davis Robbins, Dennis Robbins, Johnnie B. Robbins, Howard Roberts, Charley Romer, and A. Rush.

S - (17) - Cecil Shaddix, N. J. Shaddix, Edgar Self, Eldredge Self, Marvin Self, Arthur W. Sherbet, Morris Sherbet, William M. Sherbet, Louis Sherman, Thomas Grant Shirley, Hamp H. Sims, Chuck Smith, Harley Spradlin, William H. Stancil, Verbon Starkey, Stanly Stoddard, T F. Sullins, and Frank Sullivan.

T - (60) - Jerry Tanner, Charles Tidwell, James Tidwell, John Tipton, Milton,

Verlon Tipton, Joe Tuck, and Arlie Tollerson.

V - (2) - Edgar N. Vandegrift and Euell Vice.

W - (18) - Henry V. Wade, Mr. Waldrop, Oscar Walker, S. J. Watts, Mr. White, Alwyn White, Cliff White, Harold White, Otis White, J. H. Whited, Austin Whitley, J. C. Whitley, Herman Whittington, Odis Whittington, Taylor Whittington, Curtis Williams, Walt Williams, George Williamson, Claude Woodard, Grady C. Woodard, Ulys Woodard, and Emanuel Woods.

Y - (1) - Claude Young.

Total = 289

Mines' Production - Alabama Dept. Industrial Relation Report:

Champion 1910 - 1921, 1922 Close (none), Tait's Gap 1923 - 1924, Tait's Gap & Champion 1925 - 1928, Champion 1943- 1945, Odenville 1946 - 1948, Champion Reopened 1961 - 1968.

BRIEF HISTORY OF CHAMPION MINES
1817 - 1968
By Edward Vandegrift Gunter, Co-Owner and Miner
*(Grandson of Superintendent Ed Vandegrift and Son of
Assistant Superintendent Howard C. Gunter of the Champion Mines)*

WASHER COMPLEX – 1962

SECTION ONE

PIONEER PERIOD
In the South the Past is never past. – William Faulkner

Discovery of the rich brown ore deposits near present day Oneonta is credited to two brothers in 1817, Gabriel and John Hanby. In 1817, after serving with General Andrew Jackson as blacksmiths in the War of 1812, the two brothers from Virginia settled. John built an iron forge on Turkey Creek near Mt. Pinson, then in Blount County, and Gabriel built a three-story log inn and tavern in Blount County on the Tennessee Road at the crossing of the Little Warrior River at present day Locust Fork; he turned to trade and politics as his livelihood rather than iron working. Gabriel went on to become a distinguished Alabama statesman, serving in the Constitutional Convention in Huntsville (1819), and later in the state Senate. John Hanby followed his trade of iron worker, making what the frontiersmen of the soon-to-be state required, knives, guns, farm implements, and cooking utensils.

Gabriel and John Hanby prospected around and are credited with having discovered rich deposits of brown ore two miles east of present day Oneonta. Although there is no record of iron making activity on or near this area, later named Champion, it was not due to the lack of raw materials nor waterpower. Blount County surveyor/geologist historian George Powell (1794-1872), buried in Nectar and honored with a historical marker, described the iron-making potential in his Powell's <u>History of Blount County</u>, published 1855:

> *In Murphree's Valley are some very fine beds of iron ore on vacant land, within four miles of good water power. There are a number of good mill seats, also in this region, on vacant land. Limestone, good firestone, and a good coal bed, one foot thick, are all within a half-mile of the ore beds. With all these advantages for making iron, Blount pays annually for 30,000 pounds of Tennessee iron. The coal beds I have seen are about two feet thick and of good quality; they are in the bottom of the main prong of the Locust Fork. The smith's haul coal from those beds for ten or twelve miles. It is said there is "coal on the Little Warrior in this division."*

The above referenced ore beds would be later classified as the Champion District, by the Alabama Geological Survey, and known as the Tait's Gap and Champion Mines. This district is comprised of about twelve square miles stretching from Oneonta in a northeasterly direction to just south of Altoona, in Etowah County. In July 1900, the Tennessee Company (T.C.I.) published a book of their plants and mines, giving a description of their holdings in this district, which included all minerals lands within the Champion District:

> *These mines are located in Murphrees Valley, near Oneonta and consist*

of some 7,000 acres of brown ore land, on which are located two large double-log-washers with all the necessary machinery and other equipment. These have a total capacity of 400 tons daily.

Powell's description "within four miles of good waterpower" most likely describes the Horton Mill location north of Oneonta, while the good coal bed within a half-mile of the ore beds would certainly describe the Tait's Gap site of the 1920-era washer. The coal mining operation is clearly indicated on top of Straight Mountain, near the western slope.

During the time of Powell's survey there were also water-powered mills at Clear Springs, Allgood, Rosa, Hendricks, and Reneau Creek on Straight Mountain, which would have been possible sites for iron forges and bloomeries. Mechanical power was required to supply a steady air blast to the furnace, as well as to operate a drop-hammer to forge the iron. With the smith's hauling coal 10-12 miles, probably to Mt. Pinson to the iron forges of John and David Hanby, it is plausible that the smiths also hauled "Champion ore" there by wagon, a distance of 22 miles. During the antebellum period, however, there is no record of iron making in Blount County, with the closest furnace at Tannehill (near Bessemer), which is now a state historical park.

War Between the States

During the War for Southern Independence in 1863, after Union forces occupied Tennessee, refugee A. F. McGee founded the Mt. Pinson Iron Works on Turkey Creek, two miles west of Alabama Highway 79. This small forge and foundry aided the Southern Cause by making horseshoes and implements for the Confederate Cavalry, before being destroyed in 1864 by federal invaders. It is possible that the rich iron ore from Champion District, discovered by the Hanby brothers, was used in these three iron forges in the Mt. Pinson area, which at that time was a part of Blount County.

Financial support and other inducements were offered by the Confederate Government to assist in opening and operating coal and iron mines. On account of the location of the arsenals at Selma and Mt. Vernon, most of the coal was obtained from the Cahaba field, in Bibb and Shelby counties. However, Tuscaloosa, Jefferson, and Walker counties contributed a substantial amount of coal from the Warrior field. St. Clair County produced some from the Coosa field. Nine counties were producing iron in charcoal fired furnaces: Lamar, Tuscaloosa, Jefferson, Bibb, Shelby, Talladega, Calhoun, Jackson, and Cherokee. Alabama provided more iron to the Confederate arsenals at Selma and Rome, Georgia, than all the other Southern states combined. Sixteen furnaces provided a daily capacity of 219 tons of pig Iron.

Rousseau's Raid Crosses Champion

Although there is no record of Confederate Government mining activity at Champion, one war-related event is recorded of the "march" through the property by Union General Lovell Rousseau with his 2,700 cavalry raiders on July 13, 1864.

Ordered by Gen. William. T. Sherman, the invaders were en-route to Opelika from Decatur to destroy the last remaining railroad supplying the Confederate forces defending Atlanta. Their recorded route on that date took them from bivouac at Summit, through Blountsville, across the Locust Fork at Royal Ford, down Murphree's Valley, across present-day Champion Mines property to Spout Springs Gap. Riding in a column of two abreast, the invaders would have stretched over two miles and taken over half an hour to pass through the Champion property. One can envision the troopers filling their canteens with the cool, pure water at Spout Springs as they ascended Straight Mountain through what was then known as Allgood's Gap. The enemy met no local resistance from a Home Guard, if by chance one existed, due to the overwhelming strength of the raiders, who carried two light Rodman cannons and Spencer repeating rifles.

Rousseau Takes Alaska

The Fall 1999 issue of *Alabama Heritage* states:

> *In November 1865, sixteen months after Rousseau's raid through Alabama, he resigned from the army to serve as a U.S. congressman from Kentucky, his home state. He opposed radical reconstruction policies and became so enraged at the vituperative speeches of congressmen who had not fought in the war that on one occasion he "caned" Josiah B. Grinnell of Iowa. In March 1867, he returned to the army and was dispatched by [then President] Andrew Johnson to take possession of Alaska from the Russians. Given charge of the Department of Louisiana in 1868, Rousseau died in New Orleans a year later.*

After the War

When the War Between the States ended, the iron workers found their plants destroyed and themselves destitute. There were no funds for rebuilding, or even skilled men to undertake the work of reconstruction. The collapse was complete and, according to some, permanent. However, gradually the Birmingham District iron industry began to come alive again with the infusion of investors seeking to exploit the mineral wealth tapped briefly during the war. The Oxmoor Furnace, destroyed by Federal invaders in 1865, had been rebuilt and gone into blast in 1873, financed by Prattville founder Daniel Pratt, operating as Eureka Mining Co. Pratt's young inexperienced son-in-law Henry F. DeBardeleben, was appointed superintendent and general manager. He was only able to produce 10 tons of pig iron a day from the two 25-ton/day furnaces. Pig iron that had been selling at $40 per ton fell to $8 ton. Col. DeBardeleben then resigned his position, shut down the plant, and returned to Prattville; for a short time only.

Pig Iron – A New Process

A life-saving experiment of making iron from coke was conducted at the Oxmoor furnaces on February 28, 1876, by a syndicate of desperate Birmingham industrialists, the Cooperative Experimental Coke & Iron Company. The furnaces were converted from charcoal to coke, from cold blast to hot blast, with many other improvements under the direction of iron master Levin S. Goodrich. For the first time in the history of iron-making in Alabama, coke pig iron was made, of good quality.

The key ingredient of the successful experiment was the use of the newly discovered Pratt coal seam which was converted to coke by the newly patented Shantle process from Belgium. This coal had beaten every other coal then known in the Birmingham District for coking purposes. Major Tom Peters, Mr. Pritchard, and Col. Sam Tate brought the coal to the attention of Col. J. W. Sloss, President of South and North Railroad, who felt responsible for the ill-fated extension policy of the L&N Railroad into Alabama, which then had no revenue from any source. The success of this deal between Major Peters, Col. Sloss, and Col. H.F. DeBardeleben, whose wife was then heir to Oxmoor furnaces after the death of Daniel Pratt, was the beginning of a long friendship. In 1880 it led to the development of Blount County minerals. Col. Sloss formed Sloss Furnace Co. and built City Furnace #1 in 1881, and Furnace #2, a year later. This company later evolved into the Sloss Iron & Steel Co. and is presently known as Sloss Industries. The present-day Sloss Furnaces Museum features the last operational blast furnaces at the original site of those built in 1881-1882; Champion ore was shipped there from 1889-1968.

Part of the Sloss Furnace – Birmingham, Alabama - 1881

Champion Mines – The Beginning

In 1880 former Confederates Major Thomas Peters (Quartermaster, Selma Arsenal-CSA) and Colonel James W. Sloss (President, South & North Alabama Railroad) acquired the 280-acre tract of iron-rich land from James D. Crump, in Blount County, two miles east of present day Oneonta. It was on this tract that the "mother lode" of the iron ore occurred and to where the railroad was extended in 1889, located in Section 33, township 12S, Range 2E, thus creating the town of Oneonta. In 1882, coal & iron entrepreneur, Colonel Henry Fairchild Debardeleben, purchased Major Peters' half-interest and named the property Champion, for its highest grade iron ore known anywhere at the time, 55% iron.

The Million Dollar Deal

Col. DeBardeleben gained notoriety later that year when Major Tom Peters (promoter and land agent) brokered the first "million dollar" deal on record in the Birmingham Mineral District when he sold his Pratt Coal Co. to Col. Enoch Ensley of Memphis, Tennessee. Realizing Col. Ensley had gotten the best of the deal, within six months Col. DeBardeleben offered unsuccessfully to buy it back for two million dollars. Ethyl Armes wrote that this deal caused more excitement in the state than when US Steel bought T.C.I. in 1907, at a price of $35 million. (Ref: Ethyl Armes, Coal & Iron in Alabama)

Col. Ensley, educated at the Lebanon Law School, was more or less of a sport and one of the few moneyed men of Tennessee; his father, a large planter, had left him a million dollar inheritance. He liked swapping horses better than the law and dealt with stables and stock farms having what he termed a good time. The Colonel (an honorary title) owned plantations and stock farms all around, and headed a big Memphis real estate concern, and the Memphis Gas Company. Seeking to "out-do" the Tennessee Company, he came to Birmingham and found another Tennessee brother, Major Tom Peters, telling him he was looking for a coal mine that would knock what the T.C.I. & Railroad had up in Tennessee into *a cocked hat*. Peters then introduced him to DeBardeleben and the *deal making* had begun. A friend of his, Ben Roden, later said of Ensley, "He was a worker if ever there was one, yet he told me that up to the time he came to Birmingham, he had never done a lick of work in his life. Yet he had one of the clearest brains I ever saw in a man, and a mental force and energy you don't often come across." Another friend Captain John David Hanby (grandson of Hanby who discovered Champion ore) said of Ensley, ". . . the freest man with money I ever saw. Always setting 'em up! And he never let the widows and orphans see hard times. He was as generous as the day is long." (Ref: Ethyl Armes, p 294)

Shortly after the "million dollar deal," Col. DeBardeleben, thinking he had tuberculosis, left Birmingham for Mexico and bought a sheep ranch near Loredo, where we planned to retire, thinking never to return. He signed over to his former partner, F. L. Wadsworth, as trustee, all his mineral properties and interests. After six months recuperation he discovered he did not have tuberculosis and, regaining his old vitality, returned to the Birmingham District. He then partnered with lawyer William Underwood in the firm DeBardeleben & Underwood, capitalized at $300,000, and they set about building a new blast furnace on 30 acres of land between the railroad and 1st Avenue - naming it Mary Pratt for the colonel's second little daughter.

Shortly after the Mary Pratt Furnace went into blast in 1883, Col. DeBardeleben was forced again to take a rest from the frantic pace of the Birmingham District. After a hiatus camping out on his Loredo, Mexico ranch to recuperate his tired body, which like everything else with him was swift, DeBardeleben started back to Alabama in the fall of 1885, with his brain teeming again with big schemes. He met up with Welshman David Roberts on the return trip to Alabama and soon they had raised enough capital to form the DeBardeleben Coal & Iron Co. in 1886, capitalized at two million dollars, which doubled the Col. Ensley deal of 1881, and made another sensation. Col. DeBardeleben was elected president; David Roberts, vice-president and general manager; and Andrew Adger, secretary and treasurer. The next year DeBardeleben announced plans to build a steel-making town called Bessemer at the old mining camp of Old Fort Jonesboro in a tribute to Sir Henry Bessemer, inventor of the steel-making system that revolutionized the industry. Plans were outlined for huge operations, railroads, mines, quarries, and towns, "... such as would throw in the shade any other operations so far done in Alabama," boasted the Colonel. " I was the eagle," cried DeBardeleben, ". . . and I wanted to eat all the crawfish I could - swallow up all the little fellows, and I did it!"

We learn from the above accounts that Major Thomas Peters was involved in many of the "big deal making" among the giants of the fledging iron industry: Sloss, DeBardeleben, Ensley, Milton Smith, William Underwood, Truman Aldrich and others. (Ref: Ethyl Armes)

Champion Mines

SECTION TWO

STEAM POWER ERA
Can't you hear the whistle blowing, rise up so early in the morn'....

Finding the need for more ore for his furnaces, DeBardeleben, with the help of his friend L&N Railroad President Milton H. Smith, acquired majority stock in the Oxmoor furnaces and properties, which he had once owned and thereby added another forty thousand acres of mineral properties. Under Milton H. Smith, the Birmingham Mineral Railroad (L&N-owned) was extended from Village Springs to Champion in 1889. By the end of the decade DeBardeleben now owned several furnaces and 150,000 acres of mineral properties - more coal and iron than any other concern in the South. Smith once exclaimed about his good friend, "Ah, DeBardeleben - he is the darndest man I ever knew in my life! Why, I've spent thirty millions following that man!"

Steam Shovel and Dinkey with Crews (9 men) – 1928.

Oneonta Incorporated

Champion Mine was then the northern terminus of the Birmingham Mineral Railroad. The town of Oneonta emerged two miles west of Champion, built around the railroad "Y," a triangular track arrangement necessary for reversing the heading of the steam locomotives for the return trip to Birmingham. Blount County publications such as The Heritage of Blount County, 1976, (page 82), credit a Mr. William Newbold, L&N Superintendent, Birmingham, for naming Oneonta for his hometown of Oneonta, New York. However, due to a dearth of documentation for the mysterious Mr. Newbold, the author is inclined to think it was Milton H. Smith, L&N president from 1882-1910, who named the town, since it was he who built the track to Champion and was born in western New York, near Lake Chautauqua, in the vicinity of Oneonta, New York (documented by Ethyl Armes). The official program printed for the Oneonta, Alabama Centennial Celebration in 1991 (incorporation date), also gave credit to William Newbold, of Oneonta, New York, for naming our county seat in 1889. The New York delegation of our sister city attending the celebration went back home and began searching the records for Mr. Newbold, and found not a trace of him. In the *Map of Blount Mountain Minerals*, 1893, by State Geologist A.M. Gibson, the town is labeled "Oneon**to**", rather than Oneonta, perhaps a mapmaker's error. I recall back in the '40s and '50s, local residents would shorten the name to **"Onto"**, saying, "I've got to run down to Onto to the store."

Chepultepec – Too Difficult to Pronounce and Spell

Before the railroad came through the area in 1889, the Chepultepec post office was located at present-day Oneonta and its boundaries reached from Oneonta to Armstead and from the Dumas Ford on the Blackburn Fork of the Warrior River to beyond the Burns Gap in Sand Mountain. Current research does not provide any further details on how the name Chepultepec was selected, but we do know that one businessman there was not satisfied with the name and set about to change it.

In 1915, Frank C. Cheney, owner of Cheney Lime Co. in the town, had a problem with the name of Chepultepec, claiming his customers *could not remember, much less spell Chepultepec,* and was losing repeat business as a result. He approached the postmaster, who successfully petitioned the US Post Office Department to change the name to Allgood, in honor of the postmaster, Dr. William B. Allgood.

County Seat Moved

The construction of the railroad to Champion in 1889 resulted in the moving of the county seat from Blountsville to the new town Oneonta, incorporated February 18, 1891. This was shortly after a new brick courthouse had been constructed in Blountsville in 1888, at a cost of $16,200. Oneonta had won a runoff election for selecting the county seat in October 1889, vying against the communities of Blountsville, Bangor, Blount Springs, Chepultepec, Nectar, Hoods Crossroads, Brooksville, and Anderton (now Cleveland). Blountsville, the proud county seat for 69 years had the building, but lost the title!

The vacated new courthouse in Blountsville soon housed Blount College, which operated there until it was destroyed by fire of unknown origin in 1895. The college then moved classes into the new Ninth District Agricultural School, where it shared the building with high school classes. In 1919, this building also burned; it was located on a hill near the old court square at the location of the present Burns Memorial Park. (Ref-Heritage of Blount County, p 27)

Another twist to the extension of the L&N Railroad to Champion/Oneonta is told in the story of the Clarence community (now Susan Moore) in the Heritage of Blount County, 1976. It seems that around 1889, a Dr. Joe Hendrix built an eleven-room house there to be used as a hotel, thinking the proposed railroad would come through there. Disappointed that the railroad followed another route, he sold his house and property to R. Martin and Havana Frances Scruggs Moore. When the house was built in 1889, the Birmingham Mineral Railroad was being extended to Champion and President M. H. Smith, was only interested in hauling minerals of coal, iron, and limestone to the iron furnaces in Birmingham. Not until the A. M. Gibson report on the coal seams of Blount Mountain was published in 1893, did L&N Railroad have reason to extend the road beyond the Champion Mine, which was done in 1900 to serve the coal mines of W. T. Underwood and the future town of Altoona.

In his memoirs published in 1922, Daniel B. Bailey of Tait's Gap, describes his work at Champion and Oneonta in 1889, ". . . Ellis Clowdus and myself put 100,000 board feet of lumber on the ground to help built [sic] the Blount County Courthouse at Oneonta. The most of this lumber was 2x12 and 22 feet in length, from that up to as high as 36 feet in length." Later that year Bailey worked two mule teams on the railroad between Oneonta and Champion, describing the work as "shushing," (apparently this was moving track ballast using a slip-scrape) for which he received "an even hundred dollars." When Champion Mines first started up, Bailey had the contract to haul lumber, shingles, brick, and sand (for chimneys) to build the first ten houses built in the mining camp. He then followed by building the first mud dam, using two teams which brought him six dollars per day.

T.C.I. Purchases Debardeleben's Holdings

In June 1892, Henry Debardeleben, concerned by a rumored merger of Sloss Furnace Co. with T.C.I., was persuaded to trade his vast mineral holdings (150,000 acres valued at $10 million), generally agreed to be ". . . practically the cream of the whole mineral region," including his half-interest in Champion, to the Tennessee Coal, Iron, and Railroad Co. (T.C.I.), in exchange of $8,000,000 in T.C.I. stock, which was rumored at the time to be on the verge of bankruptcy. Said Col. DeBardeleben of the deal, "I had been an eagle eating the crawfish. Now a bigger eagle than I had ever been came along and swallowed me. That's the long and short of that trade." T.C.I. was then capitalized at $18,000,000 and owned practically all the mineral lands in the Birmingham District, including 7,000 acres in the Champion District.

The T.C.I. furnaces in Birmingham became the dominant consumer of the Champion ore, with Col. J. W. Sloss content to lease his half-interest in the mine in return for a royalty of 17-1/2 cents per ton produced, with the option to receive up to 25% of the output if so desired. For business and legal considerations, mineral owners generally leased the property to a contract operator to protect their asset from civil suits or bankruptcy resulting from poor quality or quantity, accidents, or economic forces. Champion began in 1889 with J. W. Worthington & Co. as the contract miner and changed in 1923 to Shook & Fletcher Supply Co. Over the years of production, Champion ore was shipped to all the Birmingham District blast furnaces, including T.C.I., Sloss, Woodward, U.S. Pipe, and to both Republic furnaces at Gadsden and Birmingham.

Champion Production

The only production figures found prior to 1910 for Champion Mines were provided by T.C.I.: 1893 - 61,796 tons, 1894 - 73,901 tons. By that date Col. DeBardeleben had sold his half-interest in the mine to the Tennessee Company. The data also stated that the daily capacity was about 250 tons. It is my "estimated guess" that Champion produced approximately 700,000 tons prior to 1910, when annual tonnages were available. The table below gives analyses of the washed brown ore from Champion in 1908, reported in Burchard's Report of the Ores of the Birmingham District. (Ref. Bulletin No. 400, US. Geological Survey, page 169)

Table 1
Analysis of Champion Washed Brown Ore, 1908
From Tennessee Coal, Iron & Railroad Co.

	Average Dec. 08	High Shipment	Low Shipment
Iron, Fe	47.19	52.04	42.76
Silicia, SiO	12.50	5.18	18.13
Alumina, AlO	2.44	2.13	3.72
Manganese	0.72	0.84	0.60
Phosphorus	0.26	0.24	0.18
Water	7.10	6.05	7.20

(Payment made on dry tonnage, and total metallic content − the sum of iron and manganese)

With a haul of 25-50 miles and under ordinary weather conditions, well-washed brown ore will contain 7 per cent of moisture when delivered to the furnace. This moisture is not to be confused with water of combination. the water that enters into the brown ore as a constituent and which cannot be removed under a dull red heat. Normal limonite, when pure, contains 14.44% of this combined water (water of constitution) and 85.5 % of oxide of iron (equals 59.9% of metallic iron). The chemical formula is written: $2 Fe_2 O_3 \cdot 3 H_2 O$.

Railroad Extended to Altoona

There is an interesting connection with Champion Mine and the extension in 1900 of the L&N Railroad further up Murphree's Valley to develop coal reserves at what was to become Altoona. Col. DeBardeleben's associate William T. Underwood, after selling out his Mary Pratt Properties, had gone up on Blount Mountain coal fields to prove up the Bynum, Carnes, and other seams which had been outlined by A. M. Gibson, State Geologist, in 1893. Securing 3,000 acres of the best of these coals, Mr. Underwood went to see Milton H. Smith at L&N Railroad about extending the Birmingham Mineral Railroad from Champion some

Champion Mines

12 miles further. Underwood wrote of this matter as follows:

> *In the spring of 1900, I had secured control of a body of coal lands in Western Etowah and Blount Counties, and wanted to open mines. I wanted it badly, but my lands were many miles from a railroad, and I was not able to command one-third of the money needed. I preferred opening mines on that side of my property nearest to the Alabama Great Southern Railroad, and took the matter up with the Southern officials, but got no encouragement. I then went to M. H. Smith, and found no difficulty in arousing his interest in it. I remember his saying to me: 'If you have the quality and quantity of coal you think you have, I will build you a road.' I then explained that I could not raise more than one-third of the money needed for opening the mines, and I asked him if he could aid me with that. He said he did not know but he would see. He did not keep me waiting, but acted immediately. He made me haul thirty wagonloads of coal twelve miles for test purposes. He sent experts and proved the correctness of my statement as to quality.*
>
> *He then arranged with a Louisville bank to loan my company many thousands of dollars, which we were allowed to pay off from our earnings. He began building twelve miles of road for us in May 1900 and, in the following October, we were shipping coal over it. I started this business with but a few thousand dollars of my own, and within four years' time had paid about $80,000 for the land, paid off the banks, and sold the property for a very large sum, most of which money came from outside the State and remains invested in Alabama. The country through which he built the road, and its extension on to Attalla, had been almost a wilderness. The population there has now increased ten times or more, and the city is prosperous.*

The destination of the *30 wagonloads of the coal hauled 12 miles for testing*, obviously referred to Champion mine, from which the railroad would be extended. From there the 30 loads (perhaps 30 tons) could have been loaded onto rail for shipment *to* Birmingham, or anywhere, for testing. I submit that the testing was done at Champion, as by that time, 1900, the Champion mine was under management by T.C.I., and the steam-powered washer - with the boiler rated at 450 horsepower requiring several tons of coal per day. Understanding the practicality of Underwood and Milton Smith, it is logical to assume that the *30 wagonloads of test coal* was burned with obvious success in the large boiler at Champion. Regardless of where the coal was tested, Champion had a small role in the proving of the Altoona coal and the extension of the railroad by Mr. Smith. This railroad extension thereby provided transportation for development of ore and coal deposits around Tait's Gap.

Champion Mines

SECTION THREE

THE SHOOK BROTHERS TAKE CHARGE
Trained by the Old Guard of Birmingham Ironmasters

Shook & Fletcher Supply Company was founded in 1901 in Birmingham, by Pascal G. Shook and John Fletcher, as distributors of industrial equipment. James Warner Shook joined his older brother's firm in 1914, to head up the iron ore mining division, which became the leading producer of brown ore in the state. During the same year, Paschal also bought out partner John Fletcher, who then left the company. The first reference found of S&F involvement in iron ore property was in 1909, when they owned a half-interest (with the Morris family) in the old antebellum iron works of Cane Creek, near Ohatchee. In 1915, with J. W. Shook on board they bought an ore mine in Cedartown, Georgia, from the Woodstock Iron Co., of Anniston, Alabama.

The father of these two Shook brothers, Col. Alfred M. Shook, was a long-time official with T.C.I., having served in many management positions after joining the original Tennessee Company (Nashville) in 1866 after service in the War Between the States. During the war he was captured at Fort Donaldson (1862) and was a Prisoner of War until the war's end. Col. Shook played a large role in the development of the iron & steel industry both in Tennessee and the Birmingham District and was instrumental in bringing his company to Alabama in 1886. He served as general manager and vice president of T.C.I. as it ascended to dominance in the iron & steel industry in the Birmingham District. He left the company in 1901, due to a wholesale change in management voted in by a new regime of investors.

Paschal Green Shook

Paschal Green Shook was born in Nashville, Tennessee in 1872. The oldest son of Col. Shook, he spent his childhood in Tracy City, Tennessee. He attended the Winchester Normal and the Terrell School in Decherd. Learning stenography from his father's dictation, at eighteen he took a stenographic job with the Southern Iron Company (organized by Col. Shook) in Chattanooga under direct contact with Benjamin Talbot. He came to Birmingham in 1892, as stenographer for George B. McCormack, then general manager of the Tennessee Company, and did work for all the officials of T.C.I. for several years. This contact with the higher officials was invaluable in affording him specialized business knowledge and experience which he had not had the opportunity of getting while in school. In 1896, P. G. Shook was sent by President Nat Baxter, T.C.I., to various steel works in the Pittsburgh District, Cleveland, and St. Louis, to ascertain their current practices, methods, and just what percentage of their T.C.I. product was used in the furnaces there. In Shook's report of March 2, 1896, he states:

At the Homestead Works of the Carnegie Company, they are using our Alice basic pig in identically the same manner as that of their own manufacture and in about the same proportions. Our iron gives them an acceptable mixture, in that the silicon being exceptionally low, it enables them to use to advantage their off-basic or off-Bessemer irons, containing from one to one-and-one-half per cent silicon. The same remarks apply to all users of our product.

Shook's report strongly advocated the construction of open-hearth furnaces by the Tennessee Company (T.C.I.). In this year when pig iron was selling for only $6.00 per ton at the furnace, the manufacture of steel was recognized to be more than ever essential for the welfare of the entire Birmingham District. Shook had discovered that Birmingham pig iron was as good as anyone's.

James Warner Shook

James Warner Shook was born in 1876 at Tracy City, Tennessee and was a graduate of Phillips Exeter and the Massachusetts Institute of Technology (1898) with a degree in mining engineering. He was named for James C. Warner, Shook family associate and president of the Tennessee Company (T.C.I.) before it entered the Birmingham District. Warner Shook started his career as a draftsman in the blast furnaces at South Pittsburg, Tennessee and later at the Ensley works of T.C.I. During the first handling of basic iron in 1895-96, there was adopted an invention of his, a steam device for cleaning the cinder cars mechanically, which eventually came into general use in the district. J. W. Shook was hired by the North Eastern Steel Company and was sent to England for six months to organize two blast furnaces built by American engineers. Upon his return to this country he became superintendent of the T.C.I.-Alice Furnaces and then transferred to the Ensley Furnaces as general superintendent in 1902. In 1909, J. W. Shook joined the Central Iron & Coal Company at Holt, Alabama, and by 1909, held the position of vice-president and general manager. He came to Birmingham in 1914, joining forces with his brother, Paschal, at Shook & Fletcher Supply Company as vice-president and head of ore mining. (Ref: Ethyl Armes)

Shook & Fletcher Gets Involved

The first involvement of Shook & Fletcher (S&F) with the Champion mine was a sales agreement with Gulf States Steel, of Birmingham Alabama, signed June 8, 1917, whereas S&F represented T.C.I. for the sale of ore from the mines. Gulf States Steel would buy a minimum of 1000 tons and a maximum 3000 tons per month. Ore was to be crushed to a 4-in. top size and guaranteed to exceed 42% metallic iron, with a standard quality target of 50% metallic iron. The selling price was set at $3.25 per ton of 2240 lbs., delivered f.o.b. Alabama City, Alabama, with adjustments for freight rate changes and iron content variation. It is unclear during this time frame if the mining was under contract to the J. W. Worthington Co., or if T.C.I. was actually mining the Champion ore.

The Tait's Gap mining contract was taken over by Shook & Fletcher (S&F) from Thomas Worthington in September, 1921, with Worthington retaining 15% of the profits. This mine, owned by Sloss-Sheffield Steel & Iron Co., was begun in 1920, under contract with Thomas Worthington, with Edgar N. Vandegrift as superintendent. My mother Edna Vandegrift Gunter, the eldest daughter of Vandegrift, recalled the kindness and courtesy shown her, Sister Emma, and parents, by Mr. Worthington on his frequent visits to the mine in 1920-21. Traveling by train from Birmingham, he would be met by Mr. Vandegrift at the Tait's Gap station, just 200 yards from his house, and driven to the office located about 200 yards north of the Tait's Gap Baptist Church. Worthington would have lunch with the Vandegrift family and often bring treats of fruits, candy, and flowers for the girls, ages 7 and 9. In those days the L&N Railroad ran two morning passenger trains and two afternoon trains to Gadsden and back to Birmingham, so Mr. Worthington could commute with ease from his office to the mine. My mother recalled that Mr. Worthington was approaching retirement age and his son Tom, Jr. was not inclined to take over the business; therefore, he decided to sell to the Shook's. Over the next ten or so years, Mr. Worthington would visit the Vandegrift's every few years, touring the mines, and enjoying old times, as recorded in the Vandegrift diary. In his diary entry on January 5, 1933, Ed Vandegrift wrote that he attended Mr. Worthington's funeral in Birmingham.

According to S&F records in 1921, Thomas Worthington was indebted about $31,000 to Sloss-Sheffield and bankers for the Tait's Gap mine development. His lease provided for a $25,000 advance whenever that amount had been paid out in mine development. The advance was paid in monthly installments of $5,000 each and was to be repaid, plus 6% interest, by a deduction of 50-cents per ton from the monthly settlement of $2.50 per ton ore shipped by the contractor.

Ed Vandegrift Becomes Superintendent

Ed Vandegrift had been employed by Thomas Worthington in and around Birmingham for a number of years, building rail lines and mining red ore. Vandegrift, from Odenville, in St. Clair County, was the son of Benjamin and Clara Vandegrift and great-grandson of Christopher and Rebecca Vandegrift, a Cumberland Presbyterian preacher and farmer who had migrated to St. Clair County from Laurens, South Carolina, in about 1822; he founded the Liberty Presbyterian Church at Odenville. "Mr. Ed," as he was known, had two sisters and four brothers who became merchants, teachers, farmers, saw millers, and homemakers. He married Aurelia Mae Alverson of Coal City in 1910, and moved this family from their farm in Springville to the superintendent's house in the new mining camp at Tait's Gap in 1920.

This was a railroad community on the "Murphree's Valley to Ashville" road (a link of the Huntsville-Blountsville-Talladega Road) crossing the "Oneonta to Altoona" road at Battle's store. E. D. Battle and wife had built their store in 1918 in the tiny hamlet to provide the iron ore and coal miners with weekly groceries, work gloves, tobacco, or a hand-sliced bologna & hoop cheese sandwich. Mr. Battles died in 1928, but Mrs. Battles continued the business through the 1970s. It was there on many a summer's day I would walk 200 yards down the hill from the Vandegrift home to browse the merchandise before selecting a treat - usually buying a popsicle, Coca-Cola with Tom's peanuts, or a 5-cent cup of ice cream. I also marveled at the gravity flow gas pump (Gulf Oil) with the glass measurement tank on top, which was filled to the desired quantity by stroking a lever once for each gallon of gas, then drained into the customers fuel tank. Only one pump for gasoline was necessary - regular. Imagine, a convenience store in Tait's Gap in 1948 - how *cool* was that!

In the early 1920s the Tait's Gap community grew around the attractive L&N railway station, the Baptist Church, and the two stores owned by Parker and Battles. The water supply for the mining camp houses (about 20) was piped and flowed by gravity from a boxed-in spring on the S. H. Murphree property near the top of Straight Mountain. (Ref. Heritage of Blount County)

Riding the Train with My Grandmother

In about 1948, when my grandmother Vandegrift and I flagged down the passenger train at Tait's for Attalla, the "attractive station" described above in the Heritage of Blount County had long vanished, left with only a simple open-front "bus stop" type shelter. We boarded the train with the help of the conductor's step stool and sat on the hard wooden seats. To my amazement, they were reversible to face either direction, thereby facing another seat if desired, or flipping it to the direction of travel. We enjoyed a snack of cold fried chicken and leftover biscuits from breakfast as we steamed through the valley in the open-window coach; stopping at Altoona for more riders. Advised by the conductor, we closed our open windows as we approached Tumlin Gap tunnel, lest we choke on the steam, smoke, and ashes from the coal-burning steam locomotive. West of Attalla the locomotive required a water fill-up (10-15 min) at a water tank on a creek at the junction with the Guntersville branch line. Soon we rolled into Attalla Station and were greeted by Aunt Emma Linder and baby Walton for a three-day visit. The Murphree's Valley Branch of L&N had been extended from Champion in 1900 to Altoona, then to Attalla. From there it connected with the Alabama Mineral Railroad, going to Gadsden, Anniston, Ironaton, Talladega, Childersburg, Calera, then back to Birmingham, making a big circle serving iron ore, coal, and steel shipping points. I was fortunate to have made the Tait's to Attalla trip two or three times before passenger service ended.

Tait's Gap School

There being no school in Tait's in 1922, Vandegrift set about with community help and built a two-room school house for seven grades, paid for by public subscription. Those contributing funds for the school included: Ed Vandegrift, J. C. Marshall, Hiram J. Moses, L. D. Hill, L. E. Black, W. C. King, S. B. Dillard, D. M. Davenport, A. M. Green, L. A. Beasley, H. B. Beason, E. F. Franklin, A. J. Byrd, A. Rush, T. B. Johnson, Sam Lybrand, W. W. Hill, L. J. Huie, W. H. Horton, E. Armstrong, John Cuchman, W. L. Kelton, O. M. Lybrand, Laz Whited, C. Sherer, Noah Brasseale, W. H. Bynum, Shorty Rush, William. Galbreath, R. G. Kelton, Thomas Worthington, Sr. (Operator of the Tait's Gap Mine, 1920-21), and Tom Worthington, Jr. Trustees elected were Ed Vandegrift, Hiram J. Moses and J. C. Marshall. The first session held in the summer was taught by Professor and Mrs. W. M. Self. The first regular term principal was Albert Creel, followed by Zona Head, Lucinda Jones, and the Tom Jones' twin daughters, Ruby and Opal. The school house, now a residence, still stands in the curve of the road at the foot of the mountain near to the old L&N Railroad right of way. Many of the school donors named above were ore miners. (Ref. The Heritage of Blount County, 1976)

Morrison Family

Seeing the Jones' twin girls' names above reminded me of a trio of Morrison ladies named Ruby, Pearl, and Opal, neighbors to the Vandegrifts. They lived in the Ray Morrison household across the road from the Tait's Gap Baptist Church (now paved and named Byrd Lane). I don't know how they were related but most likely they were wives and sisters of the men of the household, Ray and Cary. They had nieces my age, Emily and Martha Morrison from Birmingham, who my sister Jane and I played with during summer vacation back in the late 1940's. We built pine straw huts on the hill beside the church and played in their tiny rubble-rock playhouse, measuring maybe 3'x6' inside, built by Ray or Cary. It still stands across the road from the church amidst the tangle of wisteria vines and ruins of the Morrison home, next door to the A. J. Byrd house.

Playhouse as it looked on October 5, 2011.

Champion Mines

If you've ever seen the little hamlet setting for the movie, "Fried Green Tomatoes," then you can imagine how Tait's Gap looked in the 1940's, with all the hustle and bustle and dust from ore and coal trucks loading rail cars, and trains coming and going hourly. The Vandegrift house is now restored and owned by another family.

A WASHER OVERLOOKING THE L&N RAILROAD AT CHAMPION MINES. DINKEY CARS ARE PICTURED DUMPING ORE TO BE WASHED.

Tait's Gap Mine

In a letter from Mr. J. W. Shook to Ed Vandegrift on February 1, 1923, he writes, *This will be your authority to charge $200.00 to expense and use the money for extending your school term for two more months.* This was a generous benefit to the Tait's Gap community in permitting the new school to operate another two months, which included students Edna and Emma Vandegrift, ages 11 and 9. Mae Vandegrift had home-schooled her daughters for the previous two years before the schoolhouse was built. Mrs. Vandegrift had previously earned her teachers certificate at Jacksonville State Teachers College and had taught several years in Coal City, St. Clair County.

Shook & Fletcher took over as contract miners of the Tait's Gap mine in September 1920, operating there until ceasing operations in 1945. The washer in 1920 was located on the west side of Highway 132 near the north end of the Tait's Gap Loop road. A rail spur off the main line L&N Railroad served the steam-powered ore washer. A narrow-gage (36") tramway hauled the muck from the ore pits to the washer up to one-mile distance. Steam locomotives called "dinkeys" pulled a string of 3-4 side-dump tramcars holding five cubic yards of muck. As a rule of thumb for these ore deposits, about five cubic yards of muck would wash out to one ton of ore, or about one ton of product per tramcar of muck. The non-ore reject material from the washer consisted of chert, limestone, agate, sand, and clay.

Conveyer Belt. Men picking out mud balls and rocks. Belt is between two heaters.

Dinkeys

A typical Dinkey engine at Champion was a saddle-tank #22, purchased from the Birmingham Rail & Locomotive Works on September 22, 1927, which still operates at an amusement park in Sandusky, Ohio. The rediscovery of this little engine is discussed below.

Wheels	0-4-0-T (T – saddle water tank)
Builder/date	Vulcan Iron Works, Wilkes-Barre, Pa/1922
Empty weight:	18 tons
Driver diameter	30" (4 driver wheels)
Tractive effort	7405 pounds
Operating pressure	100 pounds/sq. inch
Cylinders	11" x 16"

The steam era ended when S&F canceled their boiler insurance policy with Clarence Brice in 1947, after selling the last steam shovel. One of the steam shovels was sold to Robbins Coal Co. in 1947 for $1,000. This buyer was Davis Robbins, who had worked for S&F at Tait's Gap in 1934 and later gained great success in the coal industry with his invention and development of the Robbins rotary drill. The nostalgic deep sound of the washer steam whistles ended in 1925 with the conversion to electric power. Those mournful sounds signaled the shift changes and dinnertime. Long silenced decades before, I never heard them at Tait's or Champion. However, in the 1950s, I recall the Belcher Lumber sawmill whistle reverberating through the hills at the Adkins Mine from the nearby Green Pond mill. Adkins also had an old Dinkey locomotive whistle they blew with compressed air to signal the hoist operator when to pull the tramcar up the slope, but it paled in comparison to the Green Pond whistle. A steam whistle works like a pipe organ – the lower the frequency of the note sounded, the larger the pipe diameter and thus the more air flow rate (or steam) required. Only the larger capacity steam boilers at an ore washer or sawmill (or factory) could supply enough steam to sound the low mournful notes.

On October 24, 1930 Vandegrift noted in his diary, "Started truck on Champion haul," then on November 9 wrote, "Ordered car to ship Dinkey from Champion," and on December 2, wrote, "Dinkies at Champion out of work." To me this marked the end of the most fascinating time of ore mining – the nostalgic steam locomotive era, for which I was born too late to witness and ride the Dinkeys as did my cousins; Vernon Vandegrift was allowed to do that by his Uncle Ed, my grandfather, while cousin Ronald Vandegrift Gunter recalled riding the Dinkeys

Champion Mines

as a youngster at the Cheney Lime Plant at Greystone, where his father Louis was superintendent after his employment at Champion Mines.

However, the change ushered in the more efficient motorized truck haulage, eliminating the constant moving of the narrow gage track to the steam shovel. But the Dinkeys were yet to run another few months at Tait's Gap. The end was chronicled somewhat prophetically. After seeing the movie "Gone with the Wind" on February 15, Vandegrift wrote in his diary on February 26, 1940, "Tait's finished with dinkies," then on March 1, "Taking up track at Tait's," and on March 4, "Taking up track at Tait's and arranging to go on truck haul." The gasoline engine era began on March 11 when Vandegrift wrote, "Started Tait's on trucks." For the record, the Dinkey locomotive era ended at Champion Mines on February 26, 1940.

Engine No. 22

The fate of one of these little Champion locomotives, Vulcan Iron Works No. 22, built in 1922, has a happy ending, chugging along on the Cedar Point & Lake Erie Railroad in Sandusky, Ohio. Thanks to the internet web surfing by ore mining historian David Brewer of Oneonta, the location of the former Champion Mines dinkey has been located and photos of engineer Jack Clements and other S&F miners from around 1928 are posted at their web page, www.cplerr.com. The following account gives the history of little engine No. 22 through its mine owners.

When in October of 1922 the Wayne Coal Co. of Clay Bank, Ohio received an 0-4-0 saddle tank steam locomotive from Vulcan Iron Works for use in their coal mines, little did the company officials realize that their engine would 40 years later head end hot shot passenger trains at Sandusky, Ohio. The little engine received the number "22."

probably for use on one of their large construction projects. Afterwards Birmingham Rail and Locomotive Co. purchased No. 22, rebuilt her and in turn sold her to the Shook and Fletcher Supply Co. of Champion, Alabama on September 22, 1927. At Champion, No. 22 joined other similar engines in transporting iron ore from the open mines in open top cars.

When the ore was practically exhausted, No. 22 went to Birmingham Rail and Locomotive Works and was again overhauled. On March 19, 1941, she was sold to Standard Coated Products Corp. of Hephzibah, Georgia. For four years, No. 22 operated along with four other Vulcans pulling trains of clay to the processing plant.

In 1945 No. 22 was sold for the seventh time, this time to Merry Bros. Brick Co. in Augusta, Georgia. By 1951, diesel electric locomotives began arriving at the brickyard, forcing the withdrawal of 22 from service. Until the early sixties 22 sat rusting in Augusta. Then in 1960 a retired construction contractor, Mr. Charles A. Weber bought 22 for his own pleasure. He transported the saddle tanker to his home in Archbold, Ohio by lowboy trailer. In order to unload the engine from the van, he built a 200 foot section of track with the one end jacked up to allow the engine to be pulled from the trailer.

Dinkey before

Dinkey after 1st

On Saturday May 21, 1960 No. 22 arrived from Georgia. Many of the residents of the small Ohio town turned out to witness one of the most interesting events they had seen in a long time. Mr. Weber directed operations as the little locomotive was backed off the truck, down the jacked up rail and into her berth in the family garden. Immediately Mr. Weber began restoring her. He added and air brake system, a headlamp, and painted her. He had the boiler inspected and approved by the state.

On special occasions during the next three summers, Mr. Weber would fire up his engine, blow the whistle, and spin her wheels (he placed the locomotive on blocks), to the delight of the neighborhood children and the fascination of the parents.

In 1962 Mr. Weber decided to "get out of the train business." He offered No. 22 for sale. At the same time Mr. Roose, owner of the park and the Cleveland Browns NFL football team at the time, was looking for locomotive power for the planned CP&LE Railroad. When he learned of the handsome little saddle tanker, he made Mr. Weber an offer which was accepted. On August 25, 1962 No. 22 was loaded aboard another trailer and trucked to Port Clinton, Ohio.

When No. 22 arrived at Sam Conte's shops in Port Clinton, work began immediately to ready her for the CP&LE's initial season. Completely stripping the engine, Sam removed the saddle tank, steel cab, stack, and boiler jacket. Within a few months 22 was decked out with a Baldwin style mahogany cab and stainless steel trim. Sam also fashioned an old style balloon stack which could be fitted over 22's original stack. This stack was used between 1964 and 1966. By mid-August 1963, 22 was at the point undergoing trial runs which she passed with flying colors and thereupon entered regular service.

After "22's" Second Restoration
With red cattle-guard, red around the light, and red on the cab

Among the changes No. 22 had undergone in Sam's shops was her conversion from a coal burner to an oil-burner. The oil tank was installed in 22's new tender, which was borrowed from Mr. Roose's 1910 Lima Prairie. This locomotive was used by the Williamson and Brown Log and Timber Co. and by the Argent Lumber Co. at Hardeeville, South Carolina. After it's years of useful service had ended, it was sold to the Stone Machine Co. of Daisy Tennessee and finally to Mr. Roose.

No. 22 remained an oil burner until 1967 when she was reconverted to a coal burner. Along with her sisters on the CP&LE, she received pony wheels in the fall of 1968.

No. 22 in glorious colors – red, blue, and yellow

Champion Mines

Charcoal Blast Furnaces

Charcoal blast furnaces consumed an enormous acreage of timberland and one can imagine a barren landscape around such furnaces. For example, the Buckeye Furnace (12 tons/day pig iron) in southern Ohio operated on charcoal from 1851-1894, producing 3,000 tons of pig iron per year. The raw materials for this production annually were: 7,888 tons ore, 411 tons limestone (flux), *and 411,000 bushels (4,110 tons) charcoal-harvested from 350 acres timberland by 48 woodcutters working 120 days from Oct-April* to produce 11,500 cords of wood to make the 411,000 bushels of charcoal. With a forest replenishment period of 20-30 years, it required 6,000-10,000 acres to maintain the supply of charcoal.

My Gunter ancestors were hill country timber cutters in Talladega County near the Ironaton charcoal furnaces, built in 1884-1885 by Stephen N. Noble of Anniston. They cut and sawed timber for railroad crossties and charcoal for the two furnaces, each rated at 40 tons per day pig iron. From the data given above for timber acreage required for making charcoal, it is evident that the Cheaha Mountains around Ironaton were most likely denuded of trees. It is no wonder that my grandfather Gunter moved into Talladega to work as a carpenter building the Bemiston Mill in 1925 - all the trees had been harvested for charcoal.

Brown Ore

Brown ore is composed of iron, oxygen, and water, and in a general way might be said to be hematite (red ore) combined with water. One of the most common brown ore minerals is known as limonite ($2Fe_2O_3 \cdot 3H_2O$), and this contains 59.9 % iron, 25.7% oxygen, and 14.4% water. Limonite is considered the best of the ores of Alabama, commanding the highest prices and a ready sale. Practically all the brown ore mined occurs in irregular masses of concretionary origin in the residual clays resulting from the decomposition of limestone; as a consequence, the mining is uncertain and expensive. Before going to the furnace, most of the ore is washed and screened; this process, together with the cost of mining, makes it the most expensive of the iron ores. It is seldom used alone, but is usually mixed with the red ore in proportions determined by the quality of the iron desired. It was used alone in the charcoal furnaces before 1876 when basic pig iron was developed using coke rather than charcoal as a furnace fuel. (Ref. Geological Survey of Alabama Statistics of Mineral Production)

Consideration should be given to some of the principal impurities that are found in iron ores: first, in order to appreciate how a deposit of ore never yields the full percentage of iron that the predominant iron mineral contains; second, in order to show the relation of the ore to the operation, of the blast furnace. Deposits of iron ore are generally so closely associated with the rocks which enclose them that they are either mixed with the deposit or chemically combined with some of the other rock-forming minerals. Among the most common of these minerals are silica, or sand; lime and magnesia, derived from limestone and dolomite; alumina, a clay mineral; manganese, a metal resembling iron; and small quantities of phosphorus. The presence of varying percentages of these and other minerals tends, of course, to reduce the percentage of metallic iron that the deposit will yield when mined, and they must be removed as far as possible in mining and almost completely in the process of making iron. Therefore, it is of the utmost importance that the average chemical composition of a deposit of iron be ascertained before an attempt is made to mine it.

Brown ore containing 50% iron was selling in 1920 for $3.00 ton f.o.b. furnace, which meant the contractor paid the freight. Metallic ore weights are given in long tons, of 2,240 pounds, rather than short tons of 2,000 pounds. The price of Number Two Foundry Pig Iron f.o.b. Birmingham, Alabama was $30.00 in 1920.

A coal tipple on the L&N Railroad was within a stone's throw from where the Tait's Gap ore mine spur branched off. Coal was conveyed from the mine atop Straight Mountain and slid down a metal chute by gravity to the tipple. The dinky lines followed along the coal seam outcrop. The 1893 Gibson map of Blount Mountain minerals identifies this mining operation in the coal seam. This coal tipple was within one mile as the crow flies to the Tait's Gap washer. I would venture a guess that nowhere else in the state was ever iron ore and coal mined in such a close proximity. George Powell wrote in 1855 of a *good clean coal bed within a half-mile of the ore beds.*

Developing Tait's Gap

After the period of severe economic depression and great uncertainty in the iron industry that began in the autumn of 1920 and continued until late 1921, conditions at the beginning of 1922 were still unsettled but appeared to have passed the crisis and to be once more improving. Forecasts made early in 1922 were conservative, indicating that a more active but reduced demand for iron ore might be expected. The production of iron ore in 1922 ranked about with that of 1906.

In a memorandum from J. W. Shook in 1922, Mr. Vandegrift was directed to increase coal purchases from six tons to twelve tons daily for boiler fuel, citing his concern to maintain an adequate stockpile. Coal-fired boilers were used for pumps, steam shovels, dinkies, and washer engines. Boiler fuel was purchased at 12 tons per day from the local coal mines and delivered in quantities of four tons per load. In July-August, 1920, Alexander & Brothers delivered 56 tons at $ 6.00 per ton, whereas in January-March, 1921, they delivered 72 tons at $4.25 per ton. Later in May-June they delivered 200 tons at $4.00 per ton. Another supplier, W. W. Hill, delivered 29 tons of run-of-mine coal during January, 1921, at $2.50 per ton, and 29 tons in February at $4.00 per ton.

A note in the ledger book by the timekeeper mentions a "company team," which may suggest that S&F miners were hauling coal to the various boilers at the pumps, washer, steam shovels, Dinkeys, and washer. Delivery men noted were: Claude Bynum, Charley Romer, Ed Armstrong, W.G. Hatley, Otis Perrin, W. W. Lybrand., E. F. Brothers, John Alexander, and Oscar Chandler. Some of the men appear on S&F employee rolls at Tait's Gap and Champion in 1929.

Champion Mines

On July 19, 1922, S&F received notification of boiler insurance coverage on the new Marion steam shovel and narrow gage locomotive, from Oneonta agent Fendley-Hagood Co. The additional premium was $32.47 for the two new boilers for the policy underwritten by The Hartford Steam Boiler Inspection & Insurance Company. Hardware orders from Moore-Handley in Birmingham in 1922 included three slip scrapes, feed, hay (which indicates the use of mules for grading, Dinkey lines and roads.) Tools and fasteners for heavy timber construction also purchased included: 4 kegs wire nails, 11 kegs boat spikes, 12 kegs track spikes, 4 kegs track bolts, 300 machine bolts (size 5/8 x 14-in.), auger bits, picks, mattocks, hammers, cant hooks, lug hooks, shovels, wrenches, cross cut saws, and claw bars.

In May, 1922, after a Sloss-Sheffield auditor had questioned his April invoice, P. G. Shook replied how the price per ton should be calculated in accordance with the contract, which apparently the auditor had not read. One might say it was a "teachable moment" for Mr. Shook:

> *The contract provides a minimum price on "normal ore" of $2.50 per ton. The price of "normal ore" or base price is fixed in the contract as follows: first, by the selling price of pig iron; second, by scale of wages; third, by the quality of the ore. $2.50 per ton is the minimum price for "normal ore;" $4.00 is the maximum. The quality of the ore shipped you in the month of April under the contract gives us a premium of $0.334 per ton, on account of the metallic content being in excess of 50%. If the metallic content had been less than 50%, we would have been correspondingly penalized. In fact, if the metallic content (iron and manganese) should be less than 44% and as much as 40%, we would get only 75% of the price of "normal ore" at time of shipment. If the metallic content is below 40%, and as much as 38%, you pay freight charges only, and for all ore below 38% iron and manganese, you pay nothing.*

In another memorandum to Vandegrift, Shook wrote that he was doubling Vandegrift's dynamite order to E. I. Dupont from five cases (40 lb./case) to ten cases in order to save a dollar per case in cost. Mine invoices for 1923 show that monthly dynamite orders were often in cases, delivered by rail at Tait's Gap. The explosives were used in the ore pits to loosen the deposits for easier loading by the steam shovel and also to blast large "dornicks" into smaller sizes that could be processed through the washer.

After one year's operation under Shook & Fletcher, the Tait's Gap mine had its best production to date, shipping 156 cars in April 1923. Vandegrift received a letter of congratulation from P .G. Shook, saying:

> *I believe it is a record in this State for production of washed brown ore through a two-log washer. In fact, I do not think this tonnage has ever been approached in the South under the same conditions with the same*

equipment.

In January 1923, P. G. Shook reported the ore production at Tait's Gap to the State Auditor, for the period September 1920-November 1922, in which they operated the mine. Apparently S&F had failed to pay taxes due to the State Treasurer when they took over operations from Thomas Worthington, and an interest penalty had been assessed also. The total tons mined during that period were 81,584.84, tons, with taxes due of $2,447.54 plus interest due of $218.15. The mine was idle during July-October 1921, resulting in an average production for the 23 months of 3,547 tons per month. The overdue taxes were due to a misunderstanding by P. G. Shook that the taxes were to be paid by the property owner, Sloss-Sheffield, and not the contract miner. Being the consummate detail man it is unlikely that Mr. Shook was ever again late with the monthly tax payments.

In October 1923, Mr. Shook wrote to the fire insurance agent in Oneonta, Fendley-Hagood Co, requesting the additional coverage of $4,000 for ore jigs recently installed. This raised the coverage from the previous $18,000 for the buildings, trestles, washers, shops, and barns other than dwellings. The dwellings were covered for the amount of $9,000, and Mr. Shook had only recently declined an offer for tornado coverage of the same amount for an additional $18 per annum. A tornado did actually hit this area, although 26 years later - and this author was directly under it!

In early 1924 the rail spur to the washer, built in 1920, required replacement of 700 cross ties and repair of the trestle. L&N performed the work, charging Shook & Fletcher $320, for labor only; which was supplied by S&F. Apparently the timbers initially installed were not treated with creosote, since they decayed so rapidly.

Tornado!

On a stormy Thanksgiving Day night in 1949, my family and I experienced a tornado while visiting at the Vandegrift home, located only a few hundred yards from the 1920's era washer, and just over the hill from the Tait's Gap Baptist Church, which was totally demolished by the twister. The entire family was seated at the dining room table when the roaring sound "like a steam locomotive" came up the track from Oneonta. Having no time to take cover we heard it pass over the house, barely missing us and demolishing the church next door. Other houses next to the church, but untouched were those of A. J. Byrd and Ray Morrison. We inspected the complete collapse of the building the next morning. No other storm damage do I recall occurring in the community, so it could be said that Mr. Shook made the proper decision by declining the tornado insurance. A larger, finer brick church was soon rebuilt on the site where it now stands.

Before the parking lot was paved at the church, sister Jane and I collected arrow heads left there by the original inhabitants of Tait's Gap.

Dam Collapses

The 1920's era muck pond at the Tait's Gap washer's retainment dam collapsed on September 6, 1929. Farms downstream were inundated in the mud and clay silt from the failed sediment impoundment. The home of Taylor Whittings was a complete loss and the farms of Steve Chandler and a Mr. Buckner were heavily damaged. Ed Vandegrift, with both P. G. and J. W. Shook immediately took action in a relief effort for the victims. Vandegrift recorded five days later in his diary on September 10, "Mrs. J. R. E. Chandler dropped dead about 9 p.m., although unsubstantiated by the author of her connection with Steve Chandler, her death was most likely attributable to the aftermath of stress from the flood." As a result of the extensive damage by the flood, several farms were purchased by Shook & Fletcher, while other landowners were compensated for their losses and remained on the property. The 1920's steam-powered washer complex at Tait's Gap was located at the GPS coordinates: 33*58'29.34"N,86*23'58.39"W. Sufficient clear water to feed the washers at both Tait's and Champion was always in short supply in the summer months; sometimes the washer was idled while the muck was stockpiled. A clear water lake, called the Moses Lake, was built about one mile north of the washer and pumped to the washer via a 10" diameter cast iron pipe. The clear water became popular as a recreational site for Civitan picnics, swimming, and fishing. The GPS coordinates for the Moses Lake location are: 33*59' 12.02"N, 86*23' 14. 11"W.

Tait's Gap Washer Complex

Living at Tait's Gap

Ed Vandegrift mine records of the Thomas Worthington Company indicate 18 company houses at Tait's Gap in 1920, ranging from two rooms to six rooms, with monthly rent from $4.00 to $12. Occupants listed in 1920 were: Ed Vandegrift, C. G. Bynum, L. A. Beasley, A. J. Byrd, Ed Armstrong, J. C. Hill, Will Galbreath, Sam Lybrand, W. M. Morton, H. Beason, A. Rush, O. M. Lybrand, Buckner, and L. G. Huie. One house was designated "Batch," another "Negro boarding house," and two others with unspecified occupants.

Many ore miners from Ironaton in Talladega County migrated to the Tait's Gap mine seeking work, including Bruce, Cary, and Ray Morrison, J. L Gunter, and Taylor Price. Ironaton was a brown ore mining town in decline, located on the Alabama Mineral Railroad, and readily accessible to Tait's Gap by rail via Anniston and Attalla or, going the other direction - to Talladega, Childersburg, Calera, and Birmingham to Tait's Gap.

Families living near the Vandegrifts in company houses in the late 1920's included Bynum, Kelton, Lybrand, Byrd, Price, Morrison, Gunter, and others. The J. L. Gunter family lived two doors down the ridge from the superintendent's house and became life-long friends of the Vandegrifts, as did the next door neighbors C .G. and Mae Bynum. In 1928, the Gunter's youngest son was born there and named Ronald Vandegrift Gunter, for Ed Vandegrift. The Gunter's had a nephew in Talladega, Howard C. Gunter, who visited them and was introduced to the Vandegrift girls, and in 1937, married Edna Mae Vandegrift, my mother. My sister Jane was born in 1939, and I in 1941, and was named Edward Vandegrift Gunter. That year my father, an Auburn graduate in electrical engineering working at the Bemiston Bag Company (cotton mill testing laboratory), in Talladega, Alabama, was offered a job as assistant superintendent at the new S&F Doc Ray mine at Woodstock. We moved there in the summer of 1941 into the new superintendent's house in the mining camp. The camp consisted of 13 houses. Ed Vandegrift was the superintendent at Doc Ray, and would spend one night a week with our family while directing the operation. Many of the Champion/Tait's Gap miners moved to Doc Ray after the closing of those mines, including Lavie Mitchell, Oscar Fullenwider, J. P. Bynum, Jr., Walt Hitt, Hoyt Owen, Webster Galbreath, Taylor Price, and Dewey Clements.

At the Doc Ray mine we were next-door neighbors to Lavie and Carrie Mitchell. Lavie, a mine foreman, had previously lived on Boss Road at Champion in the 1930s, as described by Aulden Woodard's map. Carrie was a cousin of my father, having in common a great-grandfather, Josiah Gunter who settled Guntertown on the Clay-Talladega County line, in 1850. Josiah had left South Carolina to escape the secession and war talk there, only to be drafted into the Confederate Army at age 44, in 1864, when the war was lost and nearly ended.

The Mitchell family was fraught with bad luck, tragedy upon tragedy. First, their grown son lost his life a tragic accident, the circumstances not recalled. Then their new company house at Doc Ray was destroyed by fire in about 1942. While my family was away on Army duty in WWII, they moved into our superintendent's house until their house was rebuilt. However, after its completion; they were afraid to move back to the site of the fire, so elected to move next door to us, into the house recently vacated by Walt Hitt, who left with his steam shovel when it was sold off.

Then, after we returned from WW II, next-door neighbors to the Mitchells, they lost their only daughter Vera Fancher and her daughter during childbirth. Then one cold morning, building a fire in a coal heater in the warming shack at an Adkins Mine ore pit where he was foreman, Lavie threw kerosene on still-smoldering hot coals. It exploded, seriously burning him about the face and hands. Shook & Fletcher provided excellent medical care in his long rehabilitation, through many skin grafts, and eventually Mr. Mitchell returned to the job, although horribly scarred. Then finally on a cold foggy winter morning at dawn in the early 1950's, on his way to the mines, Mr. Mitchell had just pulled out from the camp houses onto the two-lane US Highway 11, when an 18-wheeler freight truck plowed into the rear of his half-ton company pickup, engulfing the truck in flames from the ruptured gasoline tank located behind the seat. Mr. Mitchell was killed instantly by the impact. My father came upon the wreck minutes later and I recall him returning home, so visibly shaken with the terrible news, to call the Highway Patrol. This was the only on-the-job fatality I recall during my father's employment with S&F from 1941- 1968. Mrs. Mitchell moved to Bessemer a few years later to be with other family, having endured so much sorrow at Doc Ray.

The Taylor Price family lived along the railroad at the foot of Straight Mountain, almost directly across the meadow from the Vandegrifts. The two Price girls, Alice and Polly, were life-long friends of Edna and Emma Vandegrift. On summer nights with windows and doors open, the melodic sounds of the Price girls' piano carried over the meadow to the Vandegrift house on the ridge behind the church. In the 1950's Alice gave my sister Jane her large collection of piano sheet music from the 1920s. Taylor was the mine machinist there and he relocated to the Doc Ray mining camp at Woodstock in 1941, where he later retired, living with his daughter Alice. I remember passing by their front porch and hearing Alice loudly reading the paper to Mr. Price, as he had poor vision and hearing. After he passed away, Alice moved to Tuscaloosa, where she worked as a dietician for the University. The machine shop where Mr. Price worked was belt driven by a line shaft powered by a steam engine in the original setup, with machinery including a shaper, drill press, hacksaw, lathe, grinder, and forge. This shop was moved from Champion to Doc Ray in 1945, then to Adkins in 1949, then back to Champion in 1963. From 1925 and later, the line shaft was powered by electric motor. The original shop is now restored, but without the machinery, which was salvaged in 1971, with the removal of all mining equipment owned by S&F. The towering drill press is now on display at the Tannehill Museum of Iron & Steel, donated by Curtis Rivers of the Rivers Equipment Company.

Shook & Fletcher Lease Champion

In July 1923 Shook & Fletcher began negotiations with T.C.I. and Sloss for the mining lease on the Champion property, which had been abandoned and most usable machinery removed. The previous contract miner on the property was the J. W. Worthington Co., which presumably began in 1889 when Col. DeBardeleben completed the rail line to Champion. T.C.I. bought out DeBardeleben's half-interest in Champion in 1892 and production figures were not found for years before T.C.I. arrived.

The Champion mine inventory of June 1923 included 38 houses, plus barns, storage buildings, pumps, boilers, pipe, rail, water tanks, mules and wagons. The houses varied in size as follows: 18 2-room, 6 3-room, 7 4-room, 5 5-room, 2 7-room, with only four of the 38 houses occupied. T.C.I. gave the present value of the houses without repairs at $11,286. Some of these houses were constructed as early as 1889, as stated earlier, by D. B. Bailey, who hauled lumber there for constructing the first ten houses. The washer structure was rotten and worthless, with the machinery worn out and the two boilers condemned by the insurance inspector. The Shooks asked Vandegrift to evaluate the inventory and select the items usable for restarting the mine. He selected all the rails, piping, two steam pumps & boilers, 200-tons coal, buildings, machine shop, school house, commissary, and 20 houses. Apparently the steam shovels and dinkies had been previously removed. Ore haulage cars listed were twenty 2-yard "Peacock Cars," in need of new bodies and apparently obsolete since the Tait's Gap tramcars at the time were 5-yard capacity. The Shooks offered T.C.I. an annual rental of $3,300, based on a total valuation of $30,000 for the equipment and houses desired.

Champion Mines

DIARY OF EDGAR NEWTON VANDEGRIFT – 1928
Activities relating to Mining, Farming, Hobbies
Transcribed by grandson – Edward Vandegrift Gunter - 2010

Foreword: In Fall of 1920, Ed Vandegrift, with wife Aurelia Mae and daughters Edna (9) and Emma (7) moved to Tait's Gap, Alabama to begin duties as Superintendent of Tait's Gap ore mines, operated by the Thomas Worthington Company. This account of his daily activities begins January 1, 1928, after Edna had presented a five-year diary (price $3) to him for Christmas 1927. Edna was then a senior at Blount County High School in Oneonta. By 1928, the mines at Champion and Tait's Gap, operated by Shook & Fletcher Supply Co., Birmingham, were in peak production after running electricity to the three washing plants after the arrival of Alabama Power in Oneonta.

Note: Weather was important in both mining and farming and usually noted by Vandegrift's (ENV's) opening entry "On the Job" for every work day; this was deleted to save space. (Notes in *italics* are comments by author)

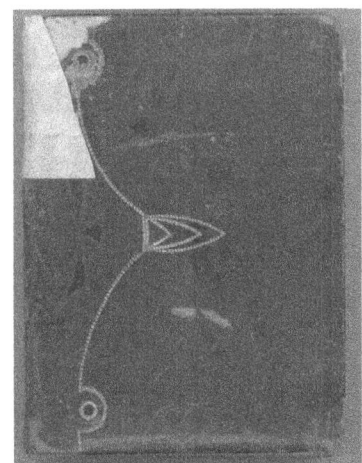

Vandergrift's 5-Year   Pages from

January 1928

1/1/28 - Left at 6:30am for hunting trip to Mississippi. Weather very cold, 10*. Arrived at Columbus about 4pm. To Mr. Sparks shortly thereafter. *(most likely a deer hunt - don't know of a Mr. Sparks))*

Jan 2 - Hunting in Mississippi. Weather so cold we came near freezing, 1° at Tait's Gap.

Jan 3 - Hunting in Miss. Weather cold, 6°, left for home at 4pm. Arrived at Bham at 10:10pm. Home at 11:45pm. Snow on ground.

Jan 4 - At home on the job. Plants all frozen up. Weather fair & cold.

Jan 5 - Pipe line trouble at Champion # 1. Also logs frozen up. Weather fair and cold, 12°. (*"logs" frozen refer to ore washer's primary cleaning vessel containing two parallel shafts, 20-30 ft long, with intermeshing steel paddles rotating in ore muck and*

separating clay from ore.)
Jan 6 - Went in AM to see damage of fire. Champion Plant started in late PM. Weather fair, 20 above.
Warming up. *(fire location not mentioned)*
Jan 7 - All plants running. Went Hunting in PM - 9 birds. Weather slightly cloudy, but pleasant.
(first mention of quail season which was to yield a bountiful harvest for ENV & guests)
Jan 9 - Mr. Shook came out. All plants working OK. Weather cloudy in AM, fair in PM rather warm about 50 above. *(ENV's boss James Warner Shook, Vice President of S&F - headed ore mining operations – brother of Paschal Green Shook, President, both sons of Alfred Montgomery Shook, Sr.)*
Jan 11 - No troubles at mines except Jones came near turning #5 shovel over. Weather fair & mild. *(steam shovels were top-heavy and could overturn if one tread got in soft ground.)*
Jan 12 - Mr. Shook came out for a Hunt. Weather fair & warm. *(refers to J. W. Shook as Mr. Shook)*
Jan 13 - Hunted a while with Mr. Hagler & Hill from Land Dept. T.C.I. Weather fair & warm. *(lands mined by S&F owned by Tennessee Coal Iron & Railroad Co. and Sloss Iron & Steel)*
Jan 14 - On the Job till 9AM. Went to Bham. Returned at 3:30. Milked for the first time in 25 years. Weather fair & warm. *(Mrs. Vandegrift, maybe away visiting, usually did milking chores)*
Jan 15 - Went to Coal City with Mr. Shook. Back home 12:35, took dinner with the Bynum's. Weather fair & warm. *(ENV & JWS (Mr. Shook) took dinner with C. G. Bynum's, next door neighbor, timekeeper. Mae Bynum- life-long friend of Mrs. Vandegrift - was a gracious hostess for the Supt. and Vice Pres. Shook)*
Jan 16 - Went Hunting a while with Mr. Strickland and party in PM. Weather fair & warm.
Jan 17 - Mr. Shook & Mr. Ryding came out for a Hunt. Weather fair & warm. *(Sloss executive)*
Jan 18 - Worked Tally out a while in PM. She did fine. Weather fair & warm. *(bird dog training, probably readying for active bird hunting season, escorting bosses & VIP's)*
Jan 20 - Went out on Bluff in PM. Hunting good. Weather fair, windy & cold.
Jan 23 - Had trouble at Pump C. Weather fair. *(probably pump at Champion)*
Jan 25 - Went Hunting in PM with Will. Got the limit. Weather fair. *(brother Will?)*
Jan 26 - Hunted with Mr. Shook. Will returned Home. Weather cloudy & mild.
Jan 28 - Went for P.R. in AM. Saw Findley & Nash. Worked on Barn in PM. Weather Cold & Windy, 20 above. *(went to town every Sat. for payroll, also saw lawyers)*
Jan 29 - Sunday. Went to #3 in AM, to Gadsden in PM. Stopped at The Pharrs (?) on the way back
Jan 31 - Mr. Shook and Mr. Shillito came out. Was in Oneonta in PM. *(Shillito was*

Champion Mines

S&F attorney)

February 1928

Feb 2 - Went Hunting with Mr. Shook, 10 birds. Weather fair & mild.

Feb 3 - Traded in 3 Chevrolets for 3 new ones. Went to Bham in PM with Lowery and Rix *(?)*. *(S&F buys new trucks -- usually trading in Oneonta)*

Feb 4 - Went for Pay Rolls in AM. Worked on Barn in PM. Weather fair & warm. Traded 1 mule to O. C. Bought 2 from Lowery. Piano tuner Mr. Thomas came out and tuned piano. *(Mr. Thomas, probably from E. F Forbes in Bham, purchased from dealer Street Lowery - could be for ENV farm)*

Feb 6 – Weather fair & warm.

Feb 8 - Went Hunting in PM, got 13. Weather windy.

Feb 9 - Went Hunting with Mr. Shook. Weather cloudy & cold.

Feb 10 - Went hunting with E.S.B. on the Bluff. Weather fair. Got Jack on 10AM train. *(Bluff is probably out on eastern edge of Straight Mt. Jack may be new bird dog)*

Feb 11 -Went Hunting in PM with Jack. 18. Weather fair & Ideal. *(Jack must be a Super Dog - did ENV shoot 18 quail!!!?)*

Feb 12 - Went to Church. Took drive with the Street Lowerys. Weather fine. Edna not feeling well. *(first mention of daughter Edna in diary)*

Feb 13 - Went Hunting in PM. 9. Weather rainy. High East Wind. *(got 9 quail)*

Feb 14 - Mr. Shook out. Went Hunting - 11. Weather fair & cool. *(got 11 quail)*

Feb 15 - Went Hunting, Tolbert settlement, 11. Weather raw, fair at 11 A.M. Rained most all PM, slow.

Feb 16 - Hunted in PM. 12. Weather fair & cool. *(got another 12 quail, often Emma would ride HB with her Dad, holding his reins as he dismounted when dog on point.)*

Feb 17 - Mr. Shook came out, went Hunting, 9. Weather rainy & cool. *(got 9 quail)*

Feb 18 - Went for P.R. in AM. At home in PM. Ordered Bat. for Buick on PM train. Weather fair. *(apparently got new battery for Buick delivered on train. Two AM and two PM trains ran through Tait's Gap to Gadsden from Bham, so purchases could be same day delivered.)*

Feb 20 - Went fairwell [sic] Hunt. Got 12. Weather fine. *(bird season ended with 12 kills. ENV had a tremendous season, reporting 114 kills from Jan 7 - Feb 20. Several hunts he had" luck" but gave no number. He mentioned bird dogs Tally and Jack, which were likely English setters.)*

Feb 22 - Weather fair. Shipped Jack to trainer in PM. *(season over - Jack returned to trainer)*

Feb 23 - Rec'd Box Cigars from Boys at Shop. Weather raining. *(ENV's 45th B 'day, so he got a gift of cigars. I never saw ENV smoke, but likely did with hunting buddies & Shooks.)*

Feb 24 - Mr. Shook came out. Carried Jim Home. Weather fair & cold. Cow got out. *(Shooks bird dog- Jim - probably boarded during season by ENV)*

Feb 25 - Went for P.R. in AM. Went to Bham in PM. Weather fair & cool.

Feb 27 - Worked on road to #3 most time. Weather cloudy & cool. *(#3 ore washer at Champion mine was probably under construction)*

Champion Mines

Feb 28 - Tuesday. Mr. Shook & Cunningham came out. Weather fair & warm. *(I don't recall any Cunningham within S&F office; this visitor may be from T.C.I. or Sloss.)*
Feb 29 - Total cars for month 443. Weather fair. *(Impressive monthly production!!!!!! Hopper cars then were 50 ton net capacity, so they shipped about 22,000 tons -more than double a good month at Champion in the 1960 s.)*

March 1928

Mar 3 -Went for Pay Roll in AM. Gardened all PM. Weather Fair.
Mar 6 - Tuesday. Sleeting in early AM. Attempted to move shovel to Champion. Backed out and shipped *(steam shovel at Tait's Gap mine - too far to crawl, so opted to put on flat-car and ship to Champion, about two miles south. These shovels had a 2-man crew; oiler, boilerman, and operator)*
Mar 7 - Wed. Loading shovel for Champion #3. Weather fair. *(apparently Champion #3 washer was about to be commissioned, so another ore pit would be required, necessitating moving machinery from Tait's Gap to Champion.)*
Mar 8 - Thurs. Mr. Shook, Regan, and Bidler came out. Weather rainy.
Mar 9 - Friday. Unloaded #2 shovel at Champion #3. Weather fair. *(took 3 days to get shovel moved, probably waiting on a flat-car and locomotive power for the special move)*
Mar 10 - Sat. Went for P.R. in AM. Rained hard most all day.
Mar 11 - At Home Sunday. Went to all dams on H.B. Water high. Weather fair. *(HB=horseback)*
Mar 12 - Monday. Water ran through S.P. Way on Minger Dam. Went to Service Club. Weather fair. *(high water overflowed through spillway - I have never heard reference of a Minger dam. Service Club later renamed Civitan Club.)*
Mar 13 - Tues. Mr. Shook & Blackburn came out. Weather fair, rained all night. *(Mr. Blackburn was probably S&F man, Wayne Blackburn, for which Russellville mine was named in 1950s, after he was killed in mine explosion.)*
Mar 15 - Thursday. Raised Bents on trestle at #3. Weather cool. *(bents are support columns of timbers in bridge/trestle construction. For washer #3 dumping)*
Mar 16 - Raised Bents on trestle. Weather Rainy.
Mar 17 - Sat. Went for P.R. in AM. Weather rainy.
Mar 21 - Mr. Shook & Laway came out. Weather fair.
Mar 22 - Thursday. Working on Trestle at C3. *(Champion #3 washer)*
Mar 24 - Saturday. Rushing track at #3. Weather rainy. *(spur track of main line)*
Mar 25 - Sunday. At House. Went for HB ride an AM. Weather rainy in PM. *(horseback)*
Mar 26 - Early. Started #2 Shovel C3. Weather fair. Mr. Shook & McGuire's came out. *(steam shovel from Tait's mine moved to Champion for new #3 washer.)*
Mar 28 - Weather cool. Mr. Fox came out. *(recall Mr. Fox as S&F purchasing?)*
Mar 30 - Bad storm during night. Several people killed near Argo. *(Argo is between Trussville and Springville- apparently this was a tornado)*

Mar 31 - Went for Pay Roll in AM. Installed lightning arrestor at Champion pump - also changed motor. *(lightning during the storm may have burned out the clear water pump)*

Vandegrift on Prince Making

April 1928

Apr 1 – Sunday. At House. Went to Church Opening. At Oneonta in AM. Went for HB. Ride in PM. Mother & Girls gone to Church tonight. *(could have been dedication day of Lester Memorial Methodist Church)*

Apr 4 - Wed. Mr. Shook, Vandegrift, and 6 Russians came out. Weather fair. *(fascinating visitors-who were they? Probably industrialists touring the "Magic City & B'ham Dist., wonder who the Vandegrift was - first name can't decipher - Visitors likely were guests of T.C.I.)*

Apr 7 - Sat. Went to Oneonta in AM for Pay Roll. Weather fair. Feeling bad in Bed in PM.

Apr 9 - Monday. On The Job. Feeling bad. Rained most all day. Very cool. Started wiring Houses at Champion. *(here a testimony of his toughness, dedication, perseverance, and leadership in reporting to work sick after weekend in bed. Company houses built by T.C.I. in 1890's were now being electrified with arrival of Alabama Power. According to account (1996) by Aulden Woodard, each house was provided with four drop lights and two 15-amp circuits. A single meter at interconnect served all S&F usage. Power was included in the house rent of $10/month. About 40 company houses remained in the 1930's according to Woodard. Wages averaged between 35-45 cents an hour for the 60-hr workweek.)*

Apr 10 - Mr. Shook and Jones came out. Weather fair.

Apr 14 - Went to town AM for P.R. Rained hard in early AM.

Apr 15 - Sunday. At home went to Church in AM. H. Back in PM. Weather cool. Having trouble with my left eye.

Champion Mines

Apr 16 - Eye sore. Weather Fair. Got pair of Mules from Lowery. *(mules were probably for mines. Street Lowery dealt in horses and mules.)*

Apr 17 - Mr. Shook came out. Weather Fair and warmer.
(presume all Mr. Shook visits were by automobile - probably driving a Buick or Packard)

Apr 19 - Wiring Houses at Champion. Weather fair.

Apr 20 - Putting up wire at Champion. Weather Fair.

Apr 21 - Up most all night. Lightning burned out switch at Champion #3. Cut in Lights at Champion. Rained all day. *(Glorious Day!!! Lights on at Champion)*

Apr 22 - Sunday. Over Works at Champion. After all night rain. Cut in Transformer at Champion.

Apr 23 - Monday. Very Heavy rain during night. Finished wiring Houses at Champion.

Apr 25 - Wed. Traded mules with Lowery. Weather fair. Ran 30 cars ore.
(first entry of daily production of ore, so he must have been pleased. At 50 tons/car, 30 cars would be 1500 tons of ore - possibly a record! Don't know if new plant #3 has gone on line yet.)

Apr 26 - Weather fair. Mr. Thomas came out to Inspect flues. *(Thomas likely from S&F Office, regarding insurance of company houses and plants fire safety.)*

Apr 28 - Went to Town for P.R. in AM. Weather Very Cool 38°.

Apr 30 - Weather fair. Ran 491 cars for Month. *(probably a new record, exceeds Feb. total of 443 cars. The electrically powered plants now are pouring out the clean ore. At 50ton/car - this month's shipment, 24,550 tons)*

May 1928

May 1 - Mr. P.G. and J.W. came out. Weather fair. *(first visit by Mr. Paschal Shook, Pres. S&F. Hope they gave ENV & crew an "atta boy" for that 491 cars shipped!!!!)*

May 4 - Moved Dinkey to dam. *(Here the Dinkey probably was needed to build up the dam)*

May 5 - Saturday. Went to town for P.R. Weather Cool.

May 7 - Monday. Started work on Dam at #3. Weather Cloudy & cool.

May 8 - Mr. Shook Came out. Weather fair and warm. Voted. Surs. Cow over Mountain. Sink Situated. Jack arrived on AM Train. *(Voted in Presidential primary for when H Hoover and Al Smith nominated. ENV probably a Democrat. Bird dog Jack must have repeated obedience school and was allowed to return to hunt again.)*

May 13 - Sunday. Went for Ride in AM. To Church for Commencement Sermon at Baptist Church. Weather fair. *(Edna's high school baccalaureate sermon)*

May 14 - Started farming. Weather fair. Rode to Champion Horseback. *(farming was second career)*

May 15 - Mr. Shook came out. Weather fair. Went to Champion at night.

May 16 - Weather fair and warm. Edna finished High School. *(Blount Co. High School)*

May 19 - Sat. Went for Pay Roll in AM. Rained Hard in PM

May 22 - Tues. Mr. Shook & Jones came out. Weather Rainy.

May 23 - Wed. Rode HB. to C. in PM. Weather fair & windy. *(rode horseback to Champion. Road ran along railroad from Tait's Gap to Champion, about 2 miles. Later used for ore haulage using Euclids in early '61 as they evaluated the Champion muck at Tait's washer)*

May 24 - Thurs. Mr. Shook, Mr. Ramsey came out. Weather fair. Went to Bessemer in late PM. Jeanette came home with us. *(could this be Erskine Ramsey, an icon of Birmingham industry, who was hired by Col. A.M. Shook as Chief Engineer, T.C.I. & Railroad, back in about 1890. Personal friend of PGS)*

May 26 - Sat. Went for P.R. in AM. Rained during night. Lightning trouble, worked all PM.

May 27 - Sunday. Went HB. Riding in AM. Car Ride in PM. Weather fair.

May 28 - Monday. Weather fair. Sink came. Worked till late putting it up. *(presented by Alabama Power Co.- sink with built-in electric dishwasher)*

May 29 - Tuesday. Shook came out. Weather fair.

May 30 - Wed. Finished installing Sink. Weather fair. *(S&F was highest power consumer in County)*

May 31 - 547 cars for month *(New monthly production record of 547 cars, about 27,000 tons - never again matched).*

June 1928

1 - Friday. Weather warm. Rained in late PM. Had Electric Stove 1 year today. *(stove was part of all-electric kitchen presented ENV's by Alabama Power Co. for S&F customer award.)*

2 - Sat. Dam trouble at #3. Went to town on train. Drove Buick Back in PM. Had tires trouble. 2 tires from Kelton. *(ENV commuted on train. Buick was in shop 2 weeks)*

4 - Monday. Dam Giving Trouble. Rained in late PM.

5 - Tues. Mr. Shook came; out. Bham Electric & Mfg. Co. man out. Worked on Champion P. Motor. Weather Rainy. *(Champion pump motor problem)*

8 - Friday. Weather fair. Mrs. O.D. Bynum, Brad (?) and Margaret Visited us. Mr. Fendley sent out an Electric Machine. *(Fendley was Singer sewing machine dealer)*

9 - Sat. Went for P.R. in AM. Fished in PM. Bought Electric Machine.

12 - Tues. Mr. Shook came out. Weather gen. Fair. Showers in spots.

15 - Sowed Hay at Moses Place. Weather warm. Fair.

16 - Sat. Went for P.R. in AM. Worked on Ball Ground at Champion in PM. *(Aulden Woodard hand-drawn map of Champion locates a Vice & Huggins ballfield)*

20 - Wed. Mr. Shook & Adkins came out. Put in switch to heater. Made application for Edna at Brenau. Weather fair & warm. *(Mr. Jess Adkins, a S&F V-P -- for whom the S&F/Caffee Junction mine near Woodstock was named in 1947, after closing Doc Ray mine)*

22 - Sowed peas in orchard in pm. #3 plant idle. Boiler trouble at shovel #2. Weather cool. Hard rain PM.

23 - Went for payroll in AM. Put in call Will in PM. Went to Oneonta in late PM.

26 - Tues. Mr. Shook came out. Listening to National Democratic Convention Keynote speech.
27 - Wed. Rain in early am. Listening in on National Dem. Convention.
30 - Sat. Went for PR in AM. At home in PM. Weather fair.

July 1928

3 - Tues. Mr. Shook came out. Weather very warm.
4 - Wed. At Home. Plants idle. Worked in gardens in am. On clothes Dinner in PM. Weather the hottest at all. 98° *(Hope it was a paid Holiday)*
7 - Sat. Went for PR in AM. On power line in PM. Mr. Worthington came out in PM. Hard rain in PM. *(Mr. Worthington had operated Tait's mines before S&F)*
8 - Sun. At Home. Drove Mr. Worthington over workings. Tom came out about 3 pm. Weather pleasant. *(Tom may be Mr. W's son, Tom, Jr.)*
13 - Fri. Sowed cabbage and turnips in PM. Weather fair. Sowed peas at Champion.
14 - Sat. Went for PR in AM. Worked in pipe line at #3 in PM. Put in shovel axle on #3 shovel.
15 - Sun. At Home. Mr. Jones came out and fished in lake. Weather fair. *(Jones was from TCI land dept.)*
19 - Thurs. Weather warm. Finished farming.
20 - Fri. Labor short on acct Decoration at Antioch. Worked in garden in late PM. No rain. *(probably refers to Antioch Methodist Church on Straight Mt.)*
21 - Sat. Went for PR in AM. Also in late PM for groceries. Hard rain in early AM. 9 pups at Reneau's *(there was a John Reneau, on baseball team)*
25 - Wed. Finished Dam. 3.25" rain fell in 1 hr. in late PM. All manner of electric Troubles. Up all Wed. night.
26 - Thur. Moved #7 from Dam to Bank. Plants down on night shift Electric trouble. Weather Rained in late PM. *(moved steam shovel out of danger from flooding)*
28 - Sat. Went for Pay Roll in AM. At home till late PM. Went to Oneonta in late PM. in Buick. Weather fair.
29 - Sun. At Home. Went for HB ride in AM to Church after Ride. Weather fair.

August 1928

4 - Sat. Went for Pay Roll in AM. At home in PM. Weather showery.
6 - Mon. Rained in late PM. Planted beans in late PM. Started on Bridge and Road changes.
10 - Fri. Went to B'ham in PM. Ran over rock and had a blowout. Rained in town but not at mines.
11 - Sat. On the Job. Went for PR in AM. Fixed tires and worked in garden all PM. Weather fair.
13 - Mon. On the Job. Went to Service Club meeting at night. Rained at night.
15 – Wed. On the Job. Put bridge in at Lowery Branch. Rained most all day. *(bridge about halfway between Tait's Gap and Oneonta)*

16 - Thur. On the Job. Mr. Shook came out. Weather fair. Irene came out in late PM.
17 - Fri. On the Job. Working on Jigs at #3. Weather fair. *(Champion washer #3)*
18 – Sat. Went for PR in AM. Working on Jig at #3 in PM. Rained in PM.
20 – Mon. On the Job. Working on Jigs at #3. *(jigs separated ore from rock by gravity)*
21 - Tues. On the Job. Mr. Shook came out. Power off 4 hrs. Rained at Tait's.
22 - Wed. On Road Vacationing. Spending night in Atlanta. Had fine day. 1 puncture. Weather fine. *(a day on road w/o a flat tire was rare)*
23 - Thur. On the Road. Atlanta to Murphry NC. Mountains Beautiful. No trouble. Weather fine. *(4-door1925 Buick coach, 6cyl-overhead valves, 191 cu in, 50hp)*
24 - Fri. On the Road. At Ashville about 2 PM. Went out Hendersonville and Lake Louise In PM. Several showers during day.
25 - Sat. On the road from Asheville to Chat. Had nice trip on Road took in Show and now returning. At 11:30 Rained several hard showers. *(Only vacation I ever heard of them taking)*
26 - Sun. On the road Chat. to Home. Went out to Fort and Mountain. Before leaving road above Attalla - Horrible. Weather Warm. *(expression "before leaving road" should be translated to mean "almost" left the road.)*
27 - Mon. On the Job. Back from trip. Everything moving satisfactory. Weather very warm. Went to Oneonta in late PM.
28 - Tues. Mr. Shook came out. Weather showery. but cooler
30 - Thur. Electrical trouble in late PM at #3. Hard rain in late PM
31 - Fri. 417 cars for month. Weather rainy. *(production slightly down from 547 car peak in May, still over 20,000 tons/mo.)*

September 1928

1 - Sat. Went for PR in AM. Bynum left for So. Ala. in AM. Went to Bham in PM. Weather Rainy most all day. *(next-door neighbor, timekeeper Clint & Mae Bynum later moved to Oneonta and became mail carrier. Children were Velma, Dalton and Margaret. They remained close family friends)*
2- Sun. At Home. Went for H.B. ride in AM. Car Hit Reaves near railroad crossing to mines. The Pharrs visited us. Rained at intervals all day.
4 - Tues. Mr. Shook came out. Weather Fair & Cool.
5 - Wed. Rode H. back to Champion in PM. Weather cloudy & cool.
6 - Thur. Rode to Champion on horse in PM. Weather fair in PM.
7 – Fri. Weather cool and fair. Working on Dam.
8 - Sat. Went for PR in AM. At Home in PM. Weather fair & cool.
10 - Mon. Edna left for Brenau at10 am. Mother went with her to Anniston. Weather fair and cool. Mr. S came out. *(Edna Mae traveling alone on train to Gainesville Ga.)*
11 - Tues. Weather fair. Put two new casings on Chev. and truck. Mileage on Chev. 6600.
12 - Wed. On the Job. Brady, mule died about 9am. Was 30 years old. Weather

fair & Pleasant.

13 - Thur. On the Job in early AM. left for Shelby at 6 am. With Mr. Shook. Had Hard days walking. Weather fair & warm. *(Mr. Shook probably had even a "harder day" - new mine in Shelby County south of Columbiana)*

15 - Sat. Went for PR in AM. For two pups in PM. 8 weeks old. Weather fair and warm. Saw local shows in PM. *(Edna has photos of pups in her album - names Bo & Jigs, English setters, as were all ENV bird dogs)*

17 - Mon. Mr. Shook came out. Worked on Hay till late PM.

20 - Thurs. Had a letter from Edna. Aus. same date. Weather cool and fair. Pump motor at Champion burned out *(electric pump motor probably replaced in 1925 the steam engine used at the clean water reservoir. That dam remains today the oldest one at Champion, ca. 1889)*

21 - Fri. Mr. Martin and foreman came out from Cedartown. Shovel boiler on #3 giving trouble. fair and cool. *(Cedartown, GA was a brown ore mining district - visitors probably miners.)*

22 - Sat. Went for PR in AM. At home in PM.

24 - Mon. Mr. Shook and Mr. Blair came out. Weather fair & cool. Mother and Em went to the fair. *(Mr. Blair was T.C.I. ore buyer and geologist)*

28 - Fri. Rode H. Back to Champion in PM. Weather fair & warm.

30 - Sun. Went for H.Back ride in am. to Church. At Home in PM. Weather fair & cool.

October 1928

1 - Mon. Mr. Shook came out. Went to town in late PM for feed. Weather fair & cool.

2 - Tues. At #3 most all day. Weather rainy.

3 - Wed. Rode Horseback to C. in PM. Rained most all a.m.

5 - Fri. Rode to Moses Hay farm in PM. to Champion after supper.

6 - Sat. Went for PR in AM. Went to St. Clair in P.M. to Aunt Delia Earlys Funeral. *(Mae Vandegrift's mother an Early.)*

8 - Mon. Mr. S, Mr. A & Mrs. A. came out. Went to Service Club at night.

12 - Fri. Went to Ala. City in PM. Weather fair and warm. Oneonta got beat 14 to nothing. *(went to football game, Emma was then a senior at BCHS)*

13 - Sat. Went for PR in AM. At House in PM doing odd jobs. Weather fair & warm.

14 - Sun. Went for Horseback ride in AM. At House rest of day. Weather fair.

15 - Mr. Shook came out. Weather rainy.

16 - Tues. Rode to Champion H.Back in PM. Went to C. after supper. Hit Rock and tore hole in oil pan on Buick. *(rough roads!)*

17 - No events of interest. Weather fair. Part time. Some rain.

18 - Thurs. Moved #7 Shovel to Dam. Mr. Shook & Miller of Sloss Co. came out. Weather Fair. *(Sloss Co. owned 50% interest in Champion Property, going back to 1880)*

19 - Sat. Went for PR in AM. Worked at #2 in PM. *(either working at washer #2 or shovel #2-can't read)*

Champion Mines

20 - Sun. At Home. Went for Ride in AM. Singing at Church. Went to Ch. #3 to fix choke coil. Charley Moses Killed his Wife Accidently. The O. B. came by in late PM. *(one of the pastures and lakes on ENV farm was named Moses-maybe it once belonged to Charley)*

22 - Tues. Mr. S. & Mr. Abbott came out. Arm bothering me terrible. Weather cool and Fair. *(I never heard Mr. Abbott mentioned - maybe a T.C.I. or Sloss official.)*

25 - Thurs. Went to see Mr. Ellison at Altoona in PM. Arranged to prospect ore on basis of 5 cents per ton for surface rights in case he has any ore. *(Mr. Ellison must have held only surface rights and not mineral, so for each ton clean ore shipped from that property, he got $0.05)*

26 - Fri. Prospecting on Ellison Place. One good Hole. Others Poor. Tally got big cut in some way. Very bad. *(Bird Dog Tally)*

29 – Mon. Mr. Shook came out. Weather cloudy in PM. Very cool.

November 1928

3 - Sat. Went for PR in AM. to Oneonta in PM. Weather rainy. At home in late PM.

4 - Sun. At Home. Went to Oneonta in late PM. Stopped by to see the Cowden's on way back. Weather cool & cloudy.

5 - Mon. Mr. Shook came out. Weather Fair & Cool.

6 - '28 Election day. On the Job. Voted Just after Noon. Weather fair. *(think ENV would have voted Democrat in '28, so that Al Smith - Herbert Hoover were elected with 58% popular vote, and thus later credited with leading country in Depression.)*

7 - Wed. Weather rainy. Agnes Box with us tonight. Winstead just called to get off to go to grandfather Funeral.

8 - Went to Bham in PM to meet Edna. Weather rainy & cold. Mary B. went with us. *(Edna's first trip home from college - via train from Gainesville, GA)*

10 - Went for PR in AM. Worked on ROW for Dinkey line to Williams Property in PM. Think I'll make it all right. Weather cool. *(ENV concerned about leasing ROW- Right of Way)*

11 - Sunday. At Home. Went to SS AM. to Oneonta in late PM for drive. Weather fair. Radioed to late PM. *(1st mention of "radioing" since the day he got new Crosby. Since he has not mentioned election results - he must have voted for Democrat Al Smith)*

13 - Tues. On the Job. Ran Dinkey line in to Williams Property in PM.

14 - Went to Bham in Early AM for Belt. Saw Estell & Jeanette. Went to Altoona in PM. Closed up all ROW for Dinkey line. *(ENV relieved - leased all ROW for new line)*

15 - Thursday. Ran levels on Dinkey line. Weather Fair & warm.

16 - Ran levels on Dinkey line in PM. Weather warm & Fair.

17 - Sat. Went for Pay Roll in AM. Worked around the house in PM.

19 - Mon. Started grade on Dinkey Line. Weather rainy in Early AM. Fair & cool in late PM. *(I sense that he was totally absorbed in building the new line - his forte!! It involved acquiring ROW, planning, surveying, grading, laying track, building trestles,*

Champion Mines

etc.)

20 - Went Hunting with Mr. Shook & Mr. Ryding. 19 birds. 8. Weather fair & very cool. *(I would conclude that ENV shot 8 of the 19 birds)*

21 - Wed. Rode H.B. from Champion. Found 1 covey. Got 5. Bought new Ford Coupe. Weather fair. *(great shooting!! About half the covey with Browning shotgun. He had three Browning's; 20g, 16g, 12g. Sadly, they were stolen sometime in the 1970's from the home. We have photos of the '28 Ford Coupe-Landau roof.)*

22 - No events of interest. Weather fair & very windy. Went out to Mr. Whited a while. Found 4 coveys - got 10. *(more great quail hunting)*

23 - Friday. Went out to Tolbert settlement in PM. Weather Fair and not so cold. *(more hunting)*

24 - Went for PR in AM. Went Hunting in PM. Got 15 -- found five coveys. *(more hunting)*

26 - Mr. Shook came out. We hunted around Tait's G. Weather fine. *(more hunting)*

27- Tues. Axle broke on #4 shovel. Wired for New one. *(this is about the 3rd axle broken among the steam shovels during year of '28)*

28 - Went out a while with Hagler & Rich - Tenn. Co. Land Dept. men. Weather cloudy.

29 - Thurs. Thanksgiving. Went Hunting with Mr. Box. We all had dinner with him and family. Weather cloudy & warm.

December 1928

1 - Sat. Went for PR in AM. Went Hunting on Coal Mine Cuts in PM. had good luck. Got the limit. *(Hunting up on Straight Mt. near Tait's Gap)*

3 - Mon. Mr. Shook, Mr. Baun (sp?) and party came out. Weather fine. Rushing's. *(could visitors be the Russian party again? don't know what ENV meant by Rushings)*

5 - Wed. Weather very cold in Early AM. had motor trouble in late PM.

6 - Sent Motor to B'ham for repairs. Weather fair. Took a round at #3 in late PM.

7 - Friday. No Events of interest. Went out on Mtn. a while in PM. fair luck. Weather fine. *(more bird hunting)*

9 - Sun. At home all day. Went for a little ride Horseback in am.

10 - Mr. Shook came out. Served Jack to Faust. Jip *(Faust bird dog Jip bred to Jack??)*

11 - Tues. Went out to Liberty in late PM to look at Cow. Weather fair.

12 - Rode Horseback to Champion in PM. Had good luck. Weather fair & warm. *(more hunting)*

13 - Thurs. At House Most all PM cleaning up. Rained in Early AM. Tait's short on water. *(refers to clear water lake levels for running washing plants)*

14 - Mr. Shook & Mr. Shepherd Came out. Had a Nice Hunt. Weather Cloudy. rained a little in Early am. *(I've never heard of Mr. Shepherd-maybe with T.C.I.)*

15 - Sat. Went for PR in AM for Emma after lunch. Took a little Hunt in PM late. Weather fair and warm. Did Cute. Drove to Oneonta in PM. Weather fair . *(Jiggs, the bird dog puppy-ENV was pleased)*

19 - Wed. Went to Bham to Meet Edna. Got back home at 12 Midnight. Weather

fair. But threatening rain at Night. *(Mother rode train from Brenau to Birmingham)*
21 - Friday. Mr. Powers and 2 other men came out to look at #3. Mr. Shooks son Alfred and friend Garby *(?sp)* came out to Hunt. Weather fine. *(No mention of Alfred Shook, III, who I met on several occasions growing up-he became vice-president after his dad passed away)*
22 - Went for PR in AM. Gave all labor $10.00 Present each. Hunted in PM. Weather fine. Good luck. Talley retrieved five. *(was one of his new pups. Times were good for the men to get a Christmas bonus)*
23 - Sun. At Home. Went for Xmas tree in AM. Drove down to Oneonta in late PM. Weather fair.
24 - Went Hunting. On mountain. Had good luck. Weather fine.
25 - Tuesday. At Home Most All day. Went out with pups & Talley a while. Weather fine. Had fun Xmas. *(he loved his hunting dogs best of all)*
26 - Took Hunt down top of mtn. Not much luck. Weather fair and cool. Looked rainy in AM.
27 - Thurs. Everything running OK. Quite a lot of flu. Came up a thunder storm through the night. Fair Today. *(mines must have taken 3 days off for Christmas.)*
28 - Mr. Shook & Son came out to Hunt. Had fair luck. Weather warm and fair. *(J. W. and son Alfred Shook, III)*
29 - Sat. Went to office in Early AM to close Pay Roll work. Stayed in rest of the day with a Cold.
30 - Sun. At Home All Day except while delivering phone messages. Edna's friends from Cleveland called on her in PM. Weather fine. *(ENV had only phone in Tait's Gap)*
31 Mon. No events of interest. Quite a bit of flu among employees. Weather Cloudy and warm.

Champion Mines

1928 Summary
Visits by boss J.W. Shook – 52
Times ENV went bird hunting - 43

Ore Shipped in 1928 - 1929

	Feb	April	May	August	
1928	443 cars	491 cars	547 cars	417 cars	
1929	Feb	March	April	May	July
	364 cars	404 cars	389 cars	389 cars	440 cars

Champion Mines

Section Four

THE GREAT DEPRESSION YEARS
Somebody told us Wall St. fell, but we were so poor we couldn't tell.
- from a song by the group Alabama

With the Wall Street Crash in October 1929, the economy began a slow decline into the Great Depression, lasting throughout the 1930's. Mine production figures for the 1930s were not found, so I will attempt to piece together some of the miners activities from the pages of my grandfather Vandegrift's personal diary. He faithfully penned an entry almost every day, always beginning with the phrase *on the job,* if it were a workday, and a weather report of *cold & rainy,* or *hot & dry.* Suffice it to say that 1928 was Champion Mines' "Best Year," with 1929 a close second.

Mining and Living during the Depression

On January 7, 1932 Vandegrift wrote in his diary *allotting ground for employees,* which I interpret to mean farm land suitable for raising livestock and community gardens, probably at Champion.

In February 1932, the Shooks called for the loading of six cars, by hand, with lump ore at Champion for shipment to T.C.I. It is unclear if mules and scrapes with wagon haulage was available or if the miners had to go into the pits with picks, shovels, and buckets to dump it into rail cars.

There was likely a direct dumping chute structure at a height above the rail cars for manually loading the cars, as seen in period washer photos, or perhaps they were able to dump the ore onto a final conveyor belt to the loading tipple. Nevertheless, the miners met the challenge and loaded six cars (300 tons) in six days. Then one week later another order came for two carloads of lump ore, and Vandegrift wrote on February 16, *ore surely scarce,* and the men went about grubbing for another 100 tons, which this time took five days to accomplish.

Then a third time, on April 13, Mr. Shook called to start loading "dornicks" again, however, the diary does not give the tonnage ordered, only that the loading process took 14 days this time around, allowing for some rainy days. One could imagine that another 100 tons, of lumps lying on the surface, were getting *really scarce.* A dornick (locally pronounced *donnick*) is an ore mining term for a large lump of ore which must be reduced in size before shipment to a blast furnace. A dornick could be as small as a basketball and be broken up with a sledge-hammer or it could be larger than the steam shovel bucket, requiring drilling with a steam (or air) jack-hammer, loaded with dynamite, and shattered. That was the last order for "hand loaded" ore lumps by the Shooks, perhaps realizing that with ore at $3.00 per ton, the labor cost was approaching that value of the ore shipped. However, it may just have been a subsidized project to provide the miners another payday. I'm sure it was welcomed work by the miners for any reason.

During March 1932, the men had torn down the Champion Washer #2, so they had likely seen the *handwriting on the wall* about their job future after the weeks of raking the hills for the final carloads of ore. On April 20th the Vandegrift's attended the wedding of his boss J. W. Shook's only son, Alfred M. Shook, III, to Miss Jane Comer, daughter of Donald Comer, President of Avondale Mills, and granddaughter of former Alabama Gov. B. B. Comer. Mr. Alfred later ascended to president of S&F in the 1960s with the passing of his uncle Paschal, and also served as a director of the Southern Company, the electric utility.

On May 31, 1932, Vandegrift wrote in his diary of receiving a *bad letter from the office cutting 50 men.* It is not known how many, if any, miners were left on the payroll after the layoff of 50 men, but there was some farming activity involving a group of some 20 men. Uncharacteristic of Mr. Shook, he did not visit the mines in 1932 until June 1st, when he and T.C.I. land agent Jones came out.

On July 8, 1932, Vandegrift noted that he *had 22 men thinning corn at Champion,* then on the 11th, *all mules and men plowing corn at Champion,* and wrote on the 13th, *finished plowing corn at Champion and Tait's.*

Harvesting continued in the fall when Mr. Shook and T.C.I. men came out to inspect the hay crop at Champion, gave a good report, and hay baling was in progress during September 22-24. From this T.C.I. inspection, it is surmised that certain stands of hay were committed for T.C.I. use; records indicate that Vandegrift was paid for the hay. At Champion a sorghum mill was brought in to make syrup from the cane crop, and the all-night cooking of the cane "squeezings" produced an impressive 195 gallons of "sopping good" sorghum syrup. On October 14, hay baling was finished at Tait's Gap and a total of 4,196 bales were harvested. Next came the corn harvest at Champion where 26 loads were gathered, then, on November 9, 155 bushels of potatoes were dug.

There was obviously a great deal of farming activity on the Tait's Gap and Champion lease acreage during the Depression years of the 1930s, and it was apparently an independent enterprise by Ed Vandegrift, obviously with an agreement with T.C.I. and the Shooks. There is no mention in his diary of any subsidy from either party other than the assumed free use of the land for farming. Support for this conclusion comes from Vandegrift's farm ledger, showing purchases of farm equipment: tractors (2), plows, hay press (2), feed mill, mules, truck, wagon, thrasher, and a farm payroll of three to five men.

Vandegrift's farm crew was engaged with this machinery all about the region in contract plowing, planting, and harvesting crops, such as corn, beans, tomatoes, hay, potatoes and sorghum cane. Vandegrift was also raising pigs, beef, and poultry for market, and selling eggs and cream. He had a sharecropper living at his Springville farm, and was sharecropping with several local landowners; he even raised enough cotton one year to make six bales for his share. Records show that most of the considerable hay crop (4,196 bales) went to Shook & Fletcher's Shelby Mine in Shelby County and to T.C.I. mines, presumably as feed for draft animals. Much of the produce, seed, and feed were sold to the miners who were also engaged in subsistence farming during hard times. Other regular customers included eggs, feed, and hay to the Shooks; eggs and cream to the Garner Hotel; hay and feed to Street Lowery; hay, syrup, eggs, pigs, beef, and feed to the local S&F miners. Vandegrift spent the remainder of November 1932 doing volunteer Red Cross work with a Dr. Miles of Oneonta, traveling throughout Blount County, making informative talks about the program. He received 200 sacks of flour and some cloth from the Red Cross and moved the supplies to the Champion commissary for distribution to the S&F miners.

The Pink Slip

The "pink slip" eventually reached the superintendent when, on March 28, 1933, Vandegrift was called to Shook's office and told his salary would now end; he noted in his diary, "now a farmer." When his final paycheck arrived by mail on March 31, he noted, ". . . starting out to try to make living farming on own hook." (on his own initiative, or independent). This news was what he had dreaded for several years. Its anticipation had intensely prepared him to succeed in farming. In the *Appendix* is included Mr. Vandegrift's farming financial records for the years 1933-38, showing that, in spite of his diversified crops, with an abundance of

Champion Mines

farming land, he still only broke even each year.

Things soon got better quickly when on May 21, P. G. Shook ordered the mines to reopen and then a month later approved a 25% pay raise to all miners. Obviously, the iron and steel industry had rebounded and the orders for iron ore had returned at last.

The following newspaper accounts give a flavor of life in Blount County in this period.

The Southern Democrat news, August 5, 1937:

> Last night we had the privilege of hearing Brother Grady Winstead preach in our church. He is a Blount County boy in many ways having worked at Champion some years ago and having married Maud Steadham, one of our local girls. Brother Winstead graduated from Howard College (now Samford University) and is going to school in Louisville, Kentucky at present. Some Blount County Church should look ahead and secure the services of Brother Winstead for a revival next year. FRIENDLY CHURCH

Note: Vandegrift's diary of 1932 tells of having to lay off two miners, Doud and Winstead. This could be the same Rev. Grady Winstead mentioned here, who attended Howard College, then on to seminary. If so, then the layoff was a career-changing event for the betterment of Winstead. The fate of Mr. Doud, Champion foreman, is unknown to the author.

Champion Mines

DIARY OF EDGAR NEWTON VANDEGRIFT - 1932-1934

Days of the Depression

By 1932 the effects of the economic depression had reached the Birmingham Iron District and orders for ore had declined, thus causing layoffs of the miners. The best source available for mining activities during this period was again the diary of Ed Vandegrift, which gives us a profile of life in the lean years compared to the best mining years of 1928 - 1929. The following entrees for 1932 - 1934 relate to farming and mining activities and are included with my comments with hopes they are helpful. *(comments in italics by author)*

1932

1/7 - allotting ground for employees *(must have been community plots for gardens & livestock)*

1/15 - got Tait's men started cleaning up.

2/1 - making plans to load six cars by hand-Champion *(Shooks get order for 6 cars)*

2/3 - started loading ore by hand – Champion *(any paying work was good work)*

2/4,5,6,8,9 - finished loading six cars by hand *(took six days to load 6 cars)*

2/7 - killed five hogs & divided them out to men *(ENV farming)*

2/16 - started loading two cars by hand for T.C.I. - ore surely scarce

2/17,18,20,22 - finished loading ore *(took five days to manually load two cars cars)*

3/17 - went to Bham for conference at office *(to Bham S&F office)*

3/28 - tearing down #2 washer dumping box and trestle *(tramcar dumping trestle)*

4/1 - working on sawmill *(ENV may have bought sawmill)*

4/5 - tearing down #2 washer - working on sawmill *(#2 washer at Champion)*

4/7 - finished saw mill setup *(sawmill at Tait's or Champion?)*

4/10 – Sunday - went to church blue all pm *(ENV got the blues over economy)*

4/13 - Mr. Shook called to start loading dornicks – Champion *(more hand loading- but its work)*

4/14,15,20,23 - loading ore at Champion went to Comer/Shook wedding *(AMS.III wedding)*

4/18 - put in bridge on road at lower end of camp *(probably at Champion camp)*

4/23 - caught up with ore loading

4/28 - finished loading ore for T.C.I. *(took 14 days to manually load 6 cars of ore)*

5/6 - T.C.I. land agent out

5/13 - Oliver Buckner boy killed by mule *(maybe son of his lumberman neighbor)*

5/14 - new armature put in Chev roadster *(his personal car 1928 Chev.)*

5/31 - had bad letter from office, 50 cut *(had to lay off 50 S&F men)*

6/1 - Shook and Jones came out *(Shook's 1st visit since December)*

6/4 - went to Bham office for confer. about power

6/17 - thrashing co. crop *(ENV contracting thrasher)*

6/18 - put 2 tires on Packard - 18,800 miles, thrashing co. peas & vetch

6/29 - tore down muck trough at Champion *(still tearing down washer #2)*

Champion Mines

7/2 - listened to Dem. party cony. -- FDR spoke in late pm   *(election year)*
7/7 - to drive off blues - went to civitan   *(ENV feeling the blues)*
7/8 - had 22 men thinning corn at Champion. new batt for Chey. *(ENV must have hired miners)*
7/11 - all mules & men plowing corn at Champion *(farming at Champion)*
7/13 - finished plowing corn at Champion and Tait's *(lots of farming on T.C.I. land)*
7/28 - finished barn plans *(maybe the Springville barn?)*
7/29 - went to Onte *(Oneonta)* for payroll in pm *(S&F still paying some workers)*
8/2 - to Spvill to cut logs for barn *(ENV owned farm in Springville before moving to Tait's)*
8/5 - cutting & bundling logs at Spvll. Buckner carried lumber over
8/13 - to office in am. got 50 sx Red Cross flour for men *(Red Cross food)*
8/5-19 - working on barn *(new barn at ENV Springville farm)*
8/23 - Tues. Mr. Shook came out about noon
9/3 - to Spvll to finish covering barn
9/13 - Mr. Shook, Jones, Murphy came out to inspect hay. report good *(Jones with TCI)*
9/22-24 - baling hay at Champion *(ENV farming on TCI lease prop.)*
10/7-8 - sorghum mill here to make syrup-195 gallons. made all night *(great soppin')*
10/12 - finished baling hay at Tait's - 4,196 bales *(a big hay crop)*
10/22 - to Spvll to dig potatoes-125 bushels *(Springville farm of env)*
10/23 - gathered 1.5 gal chestnuts in ride over farm *(ENV's cherry hill farm)*
10/27 - had possum dinner with W.W. Cowdens *(hard times for all)*
10/28 - to onte *(Oneonta)* brought out 50 sx Red Cross flour *(more Red Cross sacks of commodities)*
11/1-4 - gathering corn at Champion - 26 loads *(serious farming at Champion)*
11/9 - gathering potatoes at Champion 155 bushels *(more serious farming at Champion)*
11/11 - Red Cross work with dr. miles, to Remlap, Compton, Blount
11/14 - Red Cross work with dr. miles-western part of co.
11/16 - Red x work - got 200 sx flour *(ENV doing a lot of Red Cross work with Dr. Miles)*
11/20 - rode hb with mother & Edna-- Sunday *(all three on horseback - Emma at college)*
11/21 - rode hb to Champion -moved supplies to commissary *(Red Cross commodities?)*
11/22 - Mr. Shook came out to hunt-good day 23 *(JWS quail hunting with ENV, 23 birds)*
11/23 - got Red X cloth *(more Red Cross commodities)*
11/25 - Alfred & friend came out to hunt - rained out *(Alfred, III was son of Mr. J.W. Shook)*
11/28 - Mon went to Garner's funeral in pm *(Mr. Garner, owner of hotel)*
12/1 - rode hb with Edna *(horseback riding with daughter)*

Champion Mines

12/18 - Mr. Shook out for hunt - good luck 21 *(JWS quail hunting with ENV, 21 birds)*

12/15 - Prince fell through culvert *(his favorite horse- fate unknown.)*

12/19 - to Bham for conf to office in am *(meeting with Shooks)*

1933

1/23 - started cutting wood on mountain *(on TCI land at Tait's)*

1/24 - Mr. Shook came out -- started loading wood *(S&F may be cutting wood for TCL)*

1/26 - started cutting wood at Champion *(on TCI land at Champion)*

1/31 - loading wood at T/G and Champion *(must have cut pulpwood)*

2/1 - salary check came -- $50 cut *(ENV gets salary cut)*

2/2 - ordered car to load wood at Champion *(shipping wood by rail)*

2/3 - HB to Champion-loaded last car for Decatur *(HB=horseback, wood to Decatur)*

3/28 - to Bham to meet Mr. Shook. Boss cut me of salary. Took charge of farm the 1st of April. *(apparently ENV had bought farm, anticipating need for new livelihood).*

3/31 - drew my last salary check. starting out to try to make a living farming on own hook. *(at age 50, ENV facing mid-life crisis; now on his own initiative - independent)*

5/21 - great event -- mines to start *(hurray - for work at last)*

5/27 - in Oneonta for payroll *(S&F payroll resumes)*

6/26 - put on 20 men at Champion *(S&F recalling miners)*

6/29 - Shooks approve 25% pay raise to all *(good news for all)*

6/30 - getting #3 ready to start *(Champion washer #3, #2 washer torn down)*

7/06 - water short at Champion

7/10 - Mr. Shook came out

7/22 - S&F raised wages

7/29 - trying to get Tait's going *(trouble with Tait's washer)*

8/01 - Mr. Shook came out. got salary check *(ENV back on payroll)*

8/2 - finished car ore at Tait's *(2nd day needed to load car)*

8/5 - had new armature put in touring car *(company car, Chev. ragtop, w/side curtains)*

8/9 - Shook came out and called for 3 shifts *(must have good orders for ore)*

8/13 - went to Bham for conference

8/14 - ordered hay press from Morton Mfg. Co.

8/18 - went on 8 hr. shift. #4 shovel broke down. washer motor at Tait's burned out.

8/22 - got new hay press

8/23 - Mr.. Shook came out.

8/25 - Champion running slack. short on water.

8/30 - sent Morrison to Shelby with some washer parts. *(S&F mine at Shelby, Ala)*

9/07 - mines running poor. All lakes at Champion full.

9/14 - Mr. Shook came out. started baling hay

9/15 - changing runner in pump at Hoods Lake. *(clear water lake for washer at Tait's)*

Champion Mines

9/20 - Frank Hudson drowned or died in Hoods Lake - fishing *(Hoods Lake on farm)*
9/21 - Mr. Shook came out. baled hay until 10 at nite.
9/23 - hauled hay late at nite. J.L Gunters moved to Champion *(from Talladega)*
9/28 - Mr. Shook out. hard rain in pm. some 50 bales of hay wet.
10/6 - finished hay crop - 2000 bales *(big hay crop)*
10/13 - finished making syrup -- 230 gals sorghum *(made 35 gals. more than yr. before)*
11/17 - Paul Stewart died *(farming neighbor)*
11/20 - Mr. Shook out to hunt - great luck-8 coveys, 24 birds *(JWS & ENV - good shooting too!)*
11/23 - hunted at Odenville, 1 covey, 9 birds   *(great shooting by ENV)*

Visits by J.W. Shook - 9

1934

1/02 - loaded 4 tons hay for Shelby *(hay for mules at S&F operated ore mine at Shelby, Alabama)*
1/03 - Tait's ran 3 cars ore *(that's excellent daily run for Tait's)*
1/06 - went for p.r. in am. head still hurting me
1/08 - EM left for Brenau at 10. Mr. Shook came out
1/09 - called up Jones about pipe prices *(may be T.C.I. land agent Jones)*
1/10 - bid on pipe for CWA park *(S&F preparing bid on water line contract?)*
1/16 - to town in am. got hogs & wire
1/17 - Mr. Shook out to hunt. Mr. Reed came out to see about pipe bid
1/18 - went to Bham to figure up CWA bid
1/19 - put in bid on CWA *(maybe county water authority? S&F bidding on pipe contract)*
1/22 - put in bid on CWA work. got order in pm
1/23 - started sowing oats
1/24 - finished sowing oats in pm
1/29 - Mr. Shook and Prince Debardeleben came out *(Prince D. would have been young man; prince was grandson of Col. Henry Debardeleben, who died in 1910.)*
1/31 - started grading track & yard for stocking ore
2/3 - got battery for touring car *(ENV nephew Vernon Vandegrift recalled riding in this open top car)*
2/5 - Mr. Shook came out to hunt. fair luck
2/6 - got 500 chicks from Ala P. farms - Notasulga
2/7 - started Lybrand boy to plowing. dam broke at Champion at 9 am
2/8 - moving shovel & dinkey to dam break
2/9 - Mr. Shook came out to look at dam break
2/10-11 - working on dam at Champion
2/12 - moved #3 shovel to pit at Tait's
2/13 - moving shovel all day

2/14 - Mr. Shook out
2/16 - started Lybrand boys plowing on Mtn. Wood got sorgum seed
2/17 - thrashed cane seed
2/19 - Woodard hauled cotton seed to town for meal. got 41 sx.
2/20 - rode HB to Champion hunted on way
2/21 - old mud dam at Champion gave way again in pm
2/22 - Mr. Shook came out. bought cow from Ed Woodard
2/26 - working on dam at Champion
3/01 - still having dam trouble at Champion
3/03 - rain all day and having dam trouble at Champion
3/07 - bought feed mill
3/09 - working on dam at Champion
3/10 - to Champion and Oneonta for payroll
3/12 - started plows on Mtn.
3/13 - plowing in Campbell's fresh ground
3/14 - finished plowing for Campbell
3/15 - Mr. Shook came out. planned to build track to Champion *(from where to Champion?)*
3/16 - plowing at Champion
3/18 - rode over dinkey line and farm in am
3/20 - sick with cold. got feed mill
3/21 - Mr. Shook & Blair came out *(Blair was ore purchaser/geologist for TCI)*
3/22 - started prospecting on Ragsdale property
3/23 - put up feed mill and tried it out.
3/24 - crushed feed on new mill
3/25 - rode over proposed dinkey line in am.
3/26 - started grading dinkey line to Cham. property *(from where? to where at Champion?)*
3/27 - ran some grades on dinkey line
3/29 - Mr. Shook came out. sold him 6 chickens
3/30 - staked off some more dinkey line
4/02 - started plowing again at Champion
4/03 - went to Huntsville in am to look at plow
4/04 - staked off bal. of dinkey line
4/06 - signed order for new car
4/07 - cutting corn land on mtn.
4/09 - planted corn & beans in pm. bureau sent seed corn
4/11 - Mr. Shook out. got new Chev. coach *(probably his personal everyday farm car)*
4/15 - Sun - let Woodard have new Chev. to go to Tarrant *(loaned new car to farm worker)*
4/16 - started plowing in kudzu patch
4/19 - went to Bham in pm to sell roosters
4/20 - moved shovel from old dam to new dam. sold 146 roosters @ 22 cents
4/21 - Campbell disking new ground
4/23 - cutting ground at Champion for planting
4/24 - three Russians came out to look at washers *(likely guests of T.C.I.)*

4/25 - started planting hay at Champion. Stewart pd. bal. on sawmill *(ENV sold sawmill)*
4/26 - planted hay & corn at Champion
4/28 - baled hay for Aly in pm
4/29 - left 4:30 to see Edna - back home at 9 pm
4/30 - started turning flat field at barn
5/01 - voted in primary at noon. rigged tractor up to plow
5/02 - Campbell & Bird. planting corn in new ground
5/03 - had meeting at nite at Dem. office *(bank directors meeting)*
5/04 - Campbell started planting cotton & velvet beans
5/05 - to Bham at midday - bought some farm equipment
5/06 - Sun- fished a while in am. caught 3. one measured 22"long, 4.5#
5/10 - Mr. Shook came. fished a while - went to Civitan at nite
5/11 - Bird planted cotton. Morrison sick
5/14 - Mr. Shook came out to fish
5/15 - finished plowing Moses place. worked on disc harrow
5/17 - planting hay on kudzu patch
5/18 - planting hay on Moses place. started plowing mtn. corn
5/19 - finished laying track to new work at Champion *(dinkey line Tait's to Champion?)*
5/20 - Sun- at Champion all am moving shovel to new work for Tait's
5/21 - trying to get everything going at new work. Tait's ran her last car
5/22 - hauled first dirt from new works to Tait's *(maybe dinkey line - Tait's to Champion)*
5/23 - Mr. Shook came out - also Jones & party. washed a little ore at Tait's
5/24 - getting Tait's started off fairly well at last
5/25 - started plowing Champion corn. planting corn at hay barn
5/28 - Mr. Shook out to fish. cutting Campbell's corn land
5/29 - Oscar Walker lost his mule. Tait's running fair *(hope mule came home)*
5/30 - side dressing corn on mtn. with ciamed (??sp)
5/31 - started cutting oats in pm
6/01 - thinning corn on mtn.
6/02 - finished cutting oats in am.
6/04 - worked all day on mud pump
6/05 - motor at Champion burned out
6/05 - thinning corn at Champion
6/06 - tracks at Tait's in bad shape
6/07 - men thinning corn at Champion. Civitan at nite
6/08 - got all oats in barn. plowing on mtn.
6/09 - planted sorghum. crushed cow feed & put out. pot. plants
6/10 - drove down to Oneonta in pm. air plane wrecked. Elizabeth Bynum killed *(daughter of Joseph Porter Bynum, Sr., sister of J.P. Bynum. Jr. - timekeeper S&F)*
6/11 - Mr. Shook out
6/12 - cleaned out chicken house & moved hens at nite

6/13 - started thrashing vetch & peas in pm
6/14 - Mr. S. Stewart died at infirmary at Bham *(could this be Mr. S. Paul Stewart?)*
6/15 - went to Mr. S. Stewarts funeral in pm
6/16 - planting hay on mtn.
6/18 - thinning corn at barn
6/19 - thrashed peas for Walker
6/20 - Mr. Shook out
6/25 - started plowing corn at hay barn
6/25 - finished planting corn at Moses place
6/26 - planted tomatoes
7/01 - put 2 new tires on Packard in am *(still owned 1929 Packard, a family car)*
7/02 - baling vetch hay
7/03 - thrasher went to Greens & Phillips settlement
7/04 - worked all day. thrash on Wadsworth *(July 4th. no holiday for a farmer)*
7/05 - thrash & hay press at Reaves
7/06 - got first check on sand royalty *(mining on ENV farm land, got royalty for ore tailings)*
7/07 - in Oneonta in pm for payroll
7/08 - went out to Rosa to see puppies *(future hunting dogs)*
7/10 - planting irish pot. at Champion
7/11 - planted pot. at Tait's. Mr. Shook out
7/12 - started laying by corn at hay barn
7/13 - started stocking ore at Champion. mashed left big toe in chicken house
7/16 - started men to plowing corn at Moses place. Emma had tonsils removed
7/18 - Mr. Shook out with Alfred's two boys. Bobby & Doug *(his bro. Alfred's sons. JWS & PGS had younger brother Alfred M. Shook, Jr., JWS had son – A. Shook III)*
7/20 - went to Civitan at noon. discussed water supply for Bham
7/21 - got winks on bingo from Foust. Woodard left for Chicago
7/24 - cut some hay on Johnson grass hill
7/25 - Mr. Shook out. hard rain & wind storm in late pm. corn all blown down.
7/26 - transplanted tomatoes
7/27 - got hay in barn. cut on wed.
7/28 - sowed sorgum for hay in pm, for experiment
7/29 - the Talladega folks here
7/30 - put men to pulling weeds in hay on mtn.
7/31 - cut night shift off at Champion *(must have had good orders prior to this)*
8/01 - Mr. Shook & Alfred out. moving shovel to gravel stock *(this Alfred was JWS'S son)*
8/02 - loading gravel for rep. at Taits *(shipping gravel to Republic - I suppose)*
8/03 - finished plowing young corn. tried thrash on shelling corn.
8/08 - Mr. Shook out
8/10 - reported for gun certificate in am *(maybe a pistol permit - carried payroll weekly)*
8/11 - planted potatoes over at Champion. sold Arnold cow

8/14 - working on old #2 washer for hay storage *(thought washer had been torn down)*
8/15 - thrash went to greens. started cutting corn on mtn in pm
8/16 - went to court in am for Woods
8/18 - worked road to farm
8/20 - Mr. Shook called to stop shipments at noon.
8/21 - #3 idle. started on silage ditch in am *(Champion #3 washer idle, #2 had been dismantled in 1932)*
8/22 - started cutting corn on mtn.
8/29 - Mr. Shook out. started cutting hay at Champion
8/30 - cutting hay a Champion
8/31 - baling hay at Champion
9/01 - baling hay at Champion - started cutting hay at Tait's
9/03 - finished baling hay at Champion - 548 bales. Morrison laid up with hand
9/04 - farm labor on silo ditch.
9/05 - started cutting hay again. working on silo
9/07 - Mr. Shook called in am. baled hay in pm. *(Shook must have sent ENV to Florida. for equipment)*
9/10 - left for Florida at noon. on train all night *(likely a company business trip)*
9/11 - arrived in Florida at 9 am. weather fair *(no destination given-maybe phosphate mines to look at equipment)*
9/12 - in Florida. left for home at 9 am *(ENV very mum on mission)*
9/13 - arrived in Bham at 12 noon. home at two. good trip. Civitan club at night
9/17 - cutting hay on Moses place. Mr. Shook came out
9/18 - baling hay at Moses place
9/20 - making hay. worked until late at nite
9/21 - making hay in cane patch.
9/22 - baled hay for Parker.
9/24 - baled hay for Moses & Jorden
9/25 - baled hay for Phillips
9/27 - Mr. Shook out
9/28 - sowed crimson clover on sink hole hill
10/2- started cutting ensilage in pm
10/3 - Tait's idle
10/8 - cut sorghum hay. gathered tomatoes in pm
10/12 - finished filling silo
10/13 - In Oneonta with payroll. Had Chev checked up
10/14 - on the road to Gainsville. arrived at 9 am. 5 hrs on the road
10/16 - cutting seed beans on mtn.
10/20 - put bale of cotton in loan. gathered load of tomatoes for Talladega
10/21 - Sun - went over farm. Bing retrieved ball on first trial
10/22 - packed tomatoes in am. Mr. Shook out.
10/23 - started thrashing soy beans in pm
10/24 - finished thrashing bean seed on mtn.

10/25 - finished getting in hay on mtn & thrashed beans at mill
10/26 - picked tomatoes and dug pot. in late pm
10/27 - sent Woodard to Springville with cover crop seed. brought yearlings back
10/29 - putting in cover seed on mtn
10/30-31 - picked tomatoes in pm
11/01 - finished picking tomatoes
11/2 - packed tomatoes in am
11/03 - finished getting up sorghum hay in am. planted vetch & oats
11/04 - Sun- rode over farm in am. killed 2 geese on Hood's Lake
11/05 - started gathering corn on flat field in am
11/06 - voted in pm. went to Bham to pay off Tus? loan
11/07 - gathering corn. shipping all ore.
11/8 - Mr. Shook came out . went to Civitan at nite
11/12 - gathering corn at Champion
11/13 - thrashed beans for Bains
11/14 - dug potatoes at Champion. corn on flat field
11/15 - digging potatoes at Tait's. bought Campbell's half of potatoes for $30.00.
11/16 - planting vetch in flat field.
11/17 - planting vetch.
11/18 - Sun-rode Blalock horse a while in am
11/19 - started feeding ensilage. finished putting in vetch.
11/20 - Mr. Shook out to hunt - fair luck for rainy day - 3 coveys, 3 birds.
11/21 - had Woodard & Hare to work on potatoes. rode to Champion.
11/22 - hunted at old #3 in pm
11/23 - sent Woodard to Springville for load of corn
11/24 - cleaned beans & thrashed most all day
11/25 - Sun-went to church in am. drove by Champion & dropped Edna in pm.
11/27 - Mr. Shook out. hunting but not so good.
11/29 - hunted a while in pm
11/30 - sent Woodard to Spgville & Odenville. hunted around mud pond in pm
12/01 - hunted with Filpo Foust at Rosa
12/02 - Sun- let Hare have coach to go to Gadsden *(loaned car to farm worker)*
12/03 - killed some doves & quail in pm. had the Wittmeier, Box, Pricketts, Cowdens, & Dr. Denton for quail dinner
12/05 - bought Campbells velvet bean hay, heifer, ani? hog. rode to Champion
12/06 - gathering yellow corn
12/07 - finished yellow corn near pump - 5 loads. Campbell left in am.
12/08 - started gathering corn at Moses place
12/09 - Howard G. here *(first mention of my father visiting Edna Mae)*
12/11 - started night shift at Champion. cold, 13f in am.
12/12 - finished gathering corn. hunted to Champion.
12/13 - gathered velvet beans. took round at Tait's, fair luck. Civitan at nite
12/14 - had banker Bains & the O.D. Bynums for quail dinner
12/16 - Sun-went to church with folks. Mr. Shook called to shut mines down

Champion Mines

12/17 - Champion stocking. Mr. Shook came out.
12/18 - killed a pig. had men pull up Hugh's beans
12/19 - Emma came home. Tait's went down
12/21 - Mr. Shook out to hunt. had men get in Campbells beans
12/22 - moved shovel to stockpile at Champion *(feeding washer from stockpiled muck)*
12/24 - got 5% bonus *(Christmas bonus)*
12/26 - mine started up. hunted with Bing and Joe in pm *(Bing and Joe bird dogs)*
12/27 - rode to Champion and hunted   *(horseback hunting birds)*
12/30 - Sun- the Bro. & Mrs. Kirby came in the pm   *(Lester Memorial United Methodist pastor)*

Mine visits by Mr. Shook in 1934 = 31

ENV farm receipts, 1934 $2925.58
ENV farm expenses   3043.45
farm net (gain/loss)   - 117.87

The above farm figures & activities are given here to illustrate the difficulty of achieving financial success farming. However, there seemed to be a bountiful supply of food in the community.

Champion Mines

Miners Listed by Crew – June 1934 - In Vandegrift's Tait's Gap Time Book

Washer – am: Jim Dickey, M. D. Blalock, H. White, B. Waldon, D. Robbins, Lossie Byrd, Collet

Washer – pm: Webb Galbreath, L. Cornelius, L. S. White, K.D. Bryant, C. Payne

Dinkey - am: R. Byrd, Dickey, B. Green, Alexander, Bynum, Lowery, D. Hare

Dinkey - pm: R. Creel, Whitley, Hathcock, F. Phillips, O. Whitt, T. Green

Shop: J. H. Morrison, W. Lybrand, Self

Track: Jno. Lowe, W. Dailey, Jess Marshall, Cary Morrison, G. Ellis, Ed Lowery T. Faukersley, J. Whitley, D. Hare, Clint Payne, O. K. Green

Steam Shovel: D. Clements

SECTION FIVE

FILES OF PASCHAL G. SHOOK
*A Tennessee hillbilly born and reared far back in the
Cumberland Mountains - P. G. Shook.*

Much of the internal workings of Shook & Fletcher is revealed by correspondence from P.G. Shook, "holding down" the office and relaying daily operations news to J.W. Shook, while on vacation usually in south Florida or near Asheville NC. They were rarely out of the office at the same time, but when Mr. P. G. was away, the memorandums were infrequent, since J. W. did not seem to share his older brother's pleasure of daily dictation. The S&F office at this time was located in Birmingham in the Brown-Marx Building. In a company letter from him, on the letterhead, Fig. 6, is a list of the many manufacturing firms represented by Shook & Fletcher.

After receiving a small book, <u>Biography of a Business,</u> on the history of T.C.I. division of U.S. Steel from T.C.I. president Art Wiebel, P. G. Shook thanked him and reminisced in a letter of July 20, 1960:

> *It is of peculiar interest to me because I am personally familiar with and acquainted with so many of the facts and situations which you describe and I, too, am a Tennessee hillbilly and was born and reared to maturity in the same hills "far back in the Cumberland Mountains of Tennessee" as was TCI.*
>
> *Colonel Colyar, who was so active in the early days in the development of the properties, and my father, married sisters and he gave my father his first job at Tracy City about 1867. I grew up at Tracy City with the company, with Colonel Colyar and father in charge of operations, with exception of periods of change of management, and followed the company to Alabama about six years after father was instrumental in bringing the company into Alabama in 1886, and was allowed to work for the company until the upheaval of 1901 with the advent of Mr. Don Bacon, at which time I involuntarily walked out.* **I often said that when Bacon got here he fired everybody in the organization immediately from Mr. Baxter, President, down to me.**

As previously stated, T.C.I., under President Nathaniel Baxter, bought out Col. DeBardeleben in 1892, and the T.C.I. board voted him vice-president after trading his empire for $8 million worth of T.C.I. stock. This was only three years after completing the railroad to Champion. Always the gambler, in early 1894, DeBardeleben went up to New York City to buy up all blocks of T.C.I. so as to gain majority control of the TCI. It was the Alabama mining King's first plunge in Wall Street; there he said, "I met my Waterloo." He came back "dead broke." Although still occupying a high official position in the company, his entire personal fortune now fed the winds together with all the dollars of those of his associates who had backed him in the mad venture. Now in the great company in which he had owned millions of dollars' worth of stock, when the gambling game was over and done, DeBardeleben had not one dollar's worth. DeBardeleben resigned from T.C.I. in late 1894, and embarked on new enterprises, including Alabama Fuel & Iron Co. and other mining enterprises in the Sheffield District. He passed from this life in 1910, and was buried in Oak Hill Cemetery in Birmingham. (Ref: Ethyl Armes)

TCI Upheaval

About the T.C.I. management upheaval of 1901, author Ethyl Armes (Ref. 2) wrote that T.C.I. president Nathaniel Baxter, Jr. resigned his position in November 1901, held since 1885, when he succeeded president James C. Warner who resigned due to poor health. Baxter had joined the Tennessee Company in 1881 when John H. Inman acquired the majority interest in the company and lured him from his job as president of the First National Bank of Nashville.

Nat Baxter, like his brothers, had been born and bred in the law. The name Baxter is a well-known name in Tennessee; the men of the families were lawyers, bankers, and judges for generations. At age 15, young Nat Baxter had enlisted as a private in Freeman's Battery of Artillery and served with General Nathan B. Forrest all over the Cumberland country.

With the election of Don Bacon as president of T.C.I. in 1901, a new syndicate known as the "Gates Group," had gained control and were making a clean sweep. Beginning with Nathaniel Baxter, Jr., they removed most of the men who had been responsible for whatever progress the company had made. The official axe pared off Col. A. M. Shook, James Bowron, G .B. McCormack, Erskine Ramsey, and many lesser lights, including P. G. Shook (at age 29). Col. Shook, by then was only 56 years of age, but he chose to retire from active participation in the industrial world. He had spent his entire career with the same company. (from A. M. Shook, Sr, book)

In his retirement from the iron & steel business world, Colonel A. M. Shook maintained residences in both Tracy City and Nashville, Tennessee. Wherever he went, Col. Shook found life at its best. He kept in touch with his old friends, such as Col. Henry F. DeBardeleben, who answered one of his letters, however belatedly, addressed to his *Dear Friend*: (emphasis by author)

> *I received your letter in due course of time. It has not been answered for the reason that I have been in the woods nearly all the while. You can't imagine the pleasure that it gave me. The sentiment expressed was about what I would have expected from you.* **You gave me full credit for that which I have worked so hard to accomplish in the past.** *Your letter has been put away with an inscription on the envelope that it is to be read when my grandsons are old enough to comprehend what such a letter means. It is an affidavit from one who has been very useful himself and a man who has known all the ups and downs that the old crew had in*

bringing this district up to its present standing. You were the first man who made every sacrifice to bring about a steel era. Had you been less industrious and vigilant in your efforts to get a steel plant started, it is more than likely that it would not be started in the South up to the present time. It is hard to put upon paper just how I feel. Your sympathy was all that I received from any member of the old bunch that we used to associate with; and it came from the heart and made an impression upon me that couldn't have been done in any other way. When I see you again I shall talk to you, being a better talker than a writer. If I can live five years longer I will realize the dreams of my early manhood, and to enjoy it along with you.

Note: On January 29, 1934, H. F. D., grandson Prince DeBardeleben, visited Champion Mines with J. W. Shook, touring the works that had begun 45 years before by Henry DeBardeleben. Perhaps he had read the letter from Col. Shook to his granddad to which the above letter refers. [Van Gunter]

Colonel Shook's youngest son, Alfred M. Shook, Jr., wrote of his father's active retirement life, traveling to many distant ports: (Ref: A.M. Shook, Sr. Family Book)

My father, with his striking appearance, brilliant mind, and vivid personality, drew many people to him. In the large reception room of the old Breakers Hotel in Palm Beach he had his own table where he piled his magazines, papers and letters. A comfortable chair and a waste basket completed the group. This was his table and his chair; not a paper nor envelope was removed by the bell-boys until placed in the basket and not a single guest would presume to sit in Colonel Shook's chair, where every day he watched the world go by. Mr. Henry M. Flagler, the pioneer builder of Florida, was a great friend of his and they spent many congenial hours together at "Whitehall," built by Mr. Flagler for his home, or at The Breakers. When Mr. Flagler's private train made its first trip over the new road to Key West, Colonel Shook was one of the guests.

Colonel and Mrs. Shook crossed the Atlantic many times. Always he was asked to be the toastmaster for the banquet given by the Captain the last night at sea. Wherever he went his charm and greatness of character was recognized. And, too, he was gentle and sympathetic, easily touched by the sorrows and troubles of others - ever ready to help!

Note: With the propensity and ease with which Paschal Shook dictated letters, as evidenced in his files at S&F, I would say he was a "chip off the old Colonel's block," and probably a good toastmaster as well. - Van Gunter

Steam Shovel for Sale

On August 18, 1938, P. G. Shook writes to J. W. that they have a buyer for an idle steam shovel, Marion 1931 Traction, at Champion. Mr. Vandegrift says he never expects to use it again. Mr. P .G. will rent it for $150 a month for two months minimum, at an option to buy at any time in ten months at $1,500, rental payments to apply on purchase price. The customer, Bill Rushton, with Powhatan, was to make a final inspection that day and was expected to take the offer. Brazeal, (probably Noah Brasseal) former runner of the shovel, would go with the machine because it meant a job for him, and Vandegrift said he would like to see him get a job. They would move the shovel to a small strip coal job about seven miles north of Champion on the road to Ashville. This would likely have placed the new coal mine on Straight Mountain.

Power Problems

On August 8th and 9th, P. G. wrote to J. W. that Alabama Power Co. informed them they could not serve the Champion village or any of the former S&F employees on the existing line, as under the rules of the Utilities Public Service Commission it is not their standard construction. It would be necessary to rebuild the line at an approximate cost of $4,500, half of which S&F would have to advance on the basis of a refund of 50% of the current monthly bill plus 6% interest. Mr. Shook advised Alabama Power that this deal was out of the question. Vandegrift then did a survey at Champion and Tait's Gap and approached the renters on the basis of $5.00 per month flat rate for each house, irrespective of the character or number of appliances, but with the understanding that only the smallest capacity lamps should be used. Shook reflected that the difficulty of the plan, even if the customers accepted it, would be to collect the money, as they are out of work and would not be able to pay their current bills.

Vandegrift found enough customers willing to accept the plan that would yield within $25 of the consumption shown by the meter reading in the last day or two. He stated that S&F could well afford to stand that on account of the provision of running the pump for the water supply. The Champion drinking water was supplied from the spring at the foot of Straight Mountain on the north side of the road through Spout Springs Gap. The spring water flowed by gravity through a 2-inch pipe about 200 yards to a concrete reservoir built below the ground level and equipped with an electric pump. The reservoir, still in place, measures about 8'x8'x5' deep, giving a capacity of about 2,400 gallons. I have measured the outflow from the spring at 6 to 8 gallons per minute, about an average 400 gal/hour. This flow was sufficient to serve the twenty or more company houses acquired by S&F from the original 39 houses, plus offices and commissary listed on the 1923 inventory when T.C.I. ceased operation.

Water Problems

Former resident of the village in the 1930s, Aulden Woodard, states in his recollections that each house was equipped with one outside water spigot; however, the school/church house had no spigot, which was unfortunate and unnecessary, as the waterline from the spring ran right along the road past the school on the way to the houses. He writes that a student would have to tote water from the nearest house to the school for drinking purposes. Woodard's map of Champion shows 42 houses and a water storage tank (#42 on the map).

Apparently S&F built many more houses than the twenty houses they committed to buy in their 1923 response to the T.C.I. inventory, a (total of 39 houses. The former site of the Champion School/Church is on the west side and adjacent to the Philadelphia Baptist Church, on Champion Road, at GPS coordinates: 33*56'11.76" N, 86*26'34.10" W.

At the extreme northeastern end of the village two structures are indicated north of Champion Road as it turns SE toward the gap. One map calls the branch "Cold Water Creek" and indicates the road crossing the creek. The construction date of the concrete reservoir is unknown, but perhaps one of these numbered items is the reservoir and the other the spring. If by chance this is correct, a steam-powered pump could have been used to supply drinking water to the camp houses before the arrival of electricity.

This water supply was reactivated by S&F in 1961, during the final mining phase, pumping drinking water to the office, shop, washer, and H. C. Gunter's residence. After the mine closed in 1968, my father Howard Gunter diligently maintained this water supply to his home, until his passing in 1985. The next year my mother, Mrs. Edna Vandegrift Gunter elected to connect to the municipal water system. This pure, cold, spring water is still flowing from beneath Straight Mountain, into the headwaters of Champion Creek at 10,000 gallons per day, awaiting some entrepreneur to bottle and sell *Champion Springs Bottled Water*.

Aulden Woodard's account of the Spout Springs pipe, originating high in the rocks on the south side of Highway US 231, describes the piping installation at about the time of the new construction of US Highway 231 in 1938, when the overpass was built over the L&N Railroad and the road raised through the gap. One could wonder if perhaps the "original spout spring" (author calls it Champion Spring) may have been the village water supply, located on the north side of the gap, since the original pioneer road was about 20 ft. below the present grade, and currently the "Champion Spring" pipe empties about five feet above the branch flowing through the gap. This branch descends through the gap off Straight Mountain and merges with a small branch flowing from the NE side of the Champion village. The houses were situated on the higher ground of the valley floor, between Straight Mountain and Red Hill, such that runoff flowed either southwest toward Oneonta, or northeast toward Spout Spring gap.

During the War Between the States, it was through this gap that on July 13, 1864, Union General Lovell Rousseau led his 2,700 cavalry troops, en-route to Ashville, Talladega, and Opelika to destroy Confederate railroads. In the adjutant's report of the raid, he identifies this gap as Allgood's Gap. If, at the time, the "Champion Spring" was "boxed in" with an open sump or trough for watering passing travelers, then it is likely the cavalrymen filled their canteens and watered their horses on that hot summer day. In the 1893 Map of Blount Mountain (now

Straight Mountain) by A. M. Gibson the presently named Spout Springs gap was then identified as Allgood's Gap.

Euclids

In a letter of February 12, 1951, Mr. A. M. Shook, III wrote his Congressman Laurie C. Battle, requesting help with a delay in delivery of the last three of five Euclid trucks purchased for the Adkins mine. The Federal Government was taking a large percentage of the Euclid Company's production to be used at the H-Bomb Plant construction. Mr. Shook explained the importance of the ore production as follows:

> *For your information, we have been mining high grade iron ore in the Birmingham area for something over forty years and are today producing approximately 3,000 tons of this ore at our different mines per day. These trucks are to be used at our Adkins Mine where we produce approximately 1,000 tons per day. This ore is going to the Tennessee Company and their need for it is great. In fact, they have been after us to see what we could do to increase our tonnage. We have been hauling this ore to our washer in a skip hoist, but the area in which we worked the skip is about exhausted, and sometime in March it will be necessary for us to haul this ore by truck to the washer, the distance from the pit to the washer being something over a mile. If we are unable to get these trucks, you can see where our production will be considerably curtailed, and as someone said the other day - in these times, gold is not the precious mineral, but iron is.*

Automobile Accident

Returning from his Adkins Mine routine visit on June 19, 1952, Mr. J. W. Shook was seriously injured in a head-on automobile accident on US Highway 11 near Buckville, south of Bessemer, breaking eleven ribs. As an eleven-year-old boy, I had on a few occasions met him at Adkins while accompanying my father on his daily schedule. Mr. Shook eventually returned to work but never fully recovered from his injuries and passed away on July 28, 1955, at the age of 78.

Pig Iron – Origin of the Term

In response to an advertisement in the <u>Wall Street Journal</u> asking for the origin of the term "pig iron," P.G. Shook, at age 89, wrote the Colorado Fuel and Iron Corp., on June 22, 1961:

> *I am prompted to give you the benefit of information on this subject given me by my father over seventy years ago, at which time he was an official of the Tennessee Coal, Iron & Railroad Co., now a division of United States Steel Corp. He said that the words "pig iron" were adopted in the earliest days of the manufacture of iron, at which time there were of course no casting machines or no metal molds and all iron was cast in sand. The "pigs" were attached to the runner in the cast house sand beds and the runner was designated as*

the "sow," so that the expression "pig iron" grew out of the resemblance of the runner and the pigs to young pigs taking nourishment from their mother.

I have not the slightest doubt that this is the correct answer as to "Why do they call it PIG IRON?" I do not think any iron is now cast in sand as it is all passed through casting machines or used direct in molten state.

Gypsum Crystal Found

On July 27, 1948, Alfred M. Shook, III, S&F general superintendent of mines, received a thank-you letter from the University of Alabama Museum of Natural History for the donation of a gypsum crystal found in the Auxford Mine-Russellville ore pit. Senior Geologist Hugh Pallister, wrote:

> *I believe that this is the first time that a clear crystal of gypsum has been found completely surrounded by brown iron ore with its many phases. The crystal and part of the surrounding iron ore now rests in the museum where it is being prepared for exhibition purposes. It is certainly a beauty and a valuable addition to the collection. Dr. Jones, Mr. Dejarnette, and I thank you and your men sincerely for this kindness and assure you that we will appreciate knowledge of any other unusual occurrences which you may observe from time to time.*

Museum curator David L. Dejarnette wrote Mr. Alfred Shook on November 29, 1948:

> *We now have the specimen on exhibit in the Museum and would like for you to come by and see it when you are in Tuscaloosa. We think it is one of the most beautiful specimens we have in the Museum.*

(The Museum is located at the University of Alabama in Tuscaloosa, Alabama in Smith Hall, on the northwestern corner of University Quadrangle.)

Memo from Paschal G. Shook to James W. Shook, reported:

> *On August 18, 1948, Odenville mine foreman Bruce Morrison, driving a company pickup truck on 1st Ave. in Birmingham, was rear-ended by a drunk driver traveling at 70 MPH and tore up truck very badly. The truck was left nearby at the Tractor & Equipment Co. office for inspection by the insurance adjuster.*

Section Five

THE FINAL MINING

Diesel Power and Heavy Media Plant

On March 11, 1949, S&F received an updated ore price schedule on their T.C.I. orders, from Arthur J. Blair, Ore Agent, as follows:

Iron Price/Unit
Per percentage of iron in the ore – price

% of Iron in the Ore	Price
40 - 44.99%	$.0750
45 - 46.99%	$.0775
47 - 51.99%	$.0800
52 - 52.99%	$.0825
53% & Up	$.0850

Prices per gross dry ton (2240 lbs.) FOB Ensley Alabama, for iron and manganese combined; i.e., for a typical 50% metallic ore shipment, the price was $4.00 per dry ton.

On July 18, 1949, Alfred Shook, III, wrote Sloss-Sheffield Steel & Iron Co. confirming the shipping of the entire output from Tait's Gap Mine to the City Furnace in Birmingham. The price of the ore was to be based on combined iron and manganese content on the dry basis per gross ton (2240 lbs.) FOB Birmingham, as follows:

Iron Price/Unit
Per percentage of iron in the ore – price

% of Iron in the Ore	Price
40 – 49.99%	$.0950
45 - 46.99%	$.0975
47 - 51.99%	$.1000
52 - 52.99%	$.1025
53% & Up	$.1050

For a typical 50% metallic ore shipment, price was $5.00 per dry ton).

Tait's Gap was revived in 1949, closed in 1961, and moved to Champion.

Champion Mines

Tait's Gap Miners on Payroll, June 11, 1952 - In Vandegrift's Tait's Gap Time Book

Ed Vandegrift, Supt., Bruce Morrison, Foreman Arthur Fulenwider, C. B. Cornelius, H. J. Whittington, Hobert Henderson, Lossie Byrd, Cecil Shaddix, Jack Clements, Jesse Fulenwider, H. V. Morton, Jim Dickie, Tom Greene, E. H. Clements, Oscar Hathcock, Charles Tidwell, N. J. Shaddix, Walt Williams, Jesse Henderson, Herb Gargus, Lonnie Cornelius, D. M. Brothers, Clarence Davenport, Ulys Woodard, Emanuel Woods, Oscar Walker, B. A. Deerman.

Tait's Gap Mine Shipments of Ore and Tailings, August 1951 to September 1952

Trucking distance from washer to railroad loading ramp - 1 mile in a 1941 Ford, 5-ton, single axle dump, truck - driver: Cecil Shaddix

One day I rode with Cecil on a round trip; loading the truck in a tunnel under ore stockpile, then down the hill from the washer and across Hwy 132 and over to the dumping ramp on the L&N sidetrack. All the while he dangled a lit Lucky Strike cigarette from one corner of his mouth. When he had the car filled to capacity (50 to 70 tons) he would climb up on the hopper car, release the hand brake and roll the car down the track. He would then repeat the process on the next empty car, dropping it to the center of the ramp so each end of the car could be filled from the dumping ramp. The track was at a slight grade to permit the cars to roll without pushing - usually! Hay straw was kept on hand to chink holes in the hopper doors if the cars were in poor condition, to prevent spillage of ore en-route. On a normal day they would load about two to three cars of ore.

Tait's Gap Mine Shipments

Date	Cars with Ore	Cars with Tailings
August 1951	36	204
September 1951	46	126
October 1951	47	165
November 1951	81	107
December 1951	40	122
January 1952	55	44
February 1952	74	75
March 1952	60	21
April 1952	72	17
May 1952	68	40
June 1952	63	59
July 1952	51	26
August 1952	50	16
September 1952	84	30
14-Month Total	827	1,052
Monthly Average	59	41,000

Note: Railcars were 50-ton capacity.
Monthly average: 59 cars ore, or about 41,000 tons of ore tailings.

Tailings are a by-product from washer wastewater, recovered from dry muck ponds; ore size less than 1/4-in. is sold to cement companies in the Birmingham area at $1.00 per ton for the fine iron ore content in the sand.

Champion Mines

Tait's Gap Miners on Payroll
June, 1956 to June 12, 1957 (working 60-hr weeks)

Ed Vandegrift - Superintendent
Bruce Morrison – Foreman
Arthur Fulenwider
Lonnie Cornelius
Oscar Walker
Hobert Henderson
C. B. Cornelius
Cecil Shaddix
Jack Clements
B. A. Deerman (off job 2/16-6/3/57, reason unknown)
Jesse Henderson
Oscar Hathcock
D.M. Brothers
Jesse Fulenwider
H. V. Morton
Herbert Gargus
Jim Dickie
Tom Green
E. H. Clements
Clarence Davenport
Lossie Byrd
Walt Williams
Fred Cornelius
Arthur Madden*
J.J. Collett*
Ollie Morton*

* Miners added since 1952 roster above
(new heavy-media plant operators were Lonnie Cornelius and Arthur Madden)

Tait's Gap Mine Shipments of Ore Tailings and Gravel, June 1956 to June 1957

S&F had quote from manufacturer for Tait's Gap heavy-media plant - $75,000

A heavy-media plant was in operation during this period. It produced a clean reject of chert, flint, and gravel, sized 3/8-in. x 3-in., resulting in a cleaner product than the jigging process previously produced. The tailings now were less rich in iron due to recovery of fine ore by jigging the minus 3/8-in. material. Note the drastic curtailment in shipments to the cement plants for tailings compared to the previous period before the heavy-media plant. I don't recall rail shipment of chert, a naturally occurring mineral used for road bases; however, a pit on the property did sell chert by the truckload. When they sold the unscreened rock by-product from the heavy media plant reject belt, it was called "gravel."

Tait's Gap Mine Shipments
1956 - 1057

Date	Cars of Ore	Cars of Chert	Cars of Tailings	Cars of Scrap Iron
June 1956	94	24		
July 1956	70	26		
August 1956	84	8		
September 1956	74	11		4
October 1956	78	22	4	1
November 1956	73	21	8	
December 1956	69	5	7	
January 1957	86	3	12	
February 1957	72		8	
March 1957	85	15	14	
April 1957	80	3		
May 1957	70	6	2	
June 1957	58	15		
13-Month Total	993	156	58	5

Monthly Average – 76 cars X 50 tons = 3,800 tons/month

Various products shipped from Tait's Gap:
1. Coarse iron ore, size 4' x 0
2. Fine iron ore, size 3/8-inch, from Remer jigs at heavy-media plant washed gravel, size 4" x 0
3. Ore tailings, size-3/8", from dried muck ponds

Tait's Gap Miners on Payroll. May 14, 1958- June 28, 1959 (60-hr Workweek)
Ed Vandergrif - superintendent; Bruce Morrison - foreman.
Arthur Fulenwider, Lonnie Cornelius, Oscar Walker, Hobart Henderson, C. B. Cornelius, Cecil Shadix, Jack Clements, Oscar Hathcock, Jessie Fulenwider, H.Y. Morton, Jim Dickie, E. H. Clements, Clarence Davenport, Walt Williams, Fred Cornelius, Arthur G. Madden, J. J. Collett, B. A. Deerman, William Beard, Curtis Williams, Howard Grissom, Alton Henderson, James Tidwell, William Carroll, D. M. Brothers, Herbert Gargus, Tom Green, Lossie Byrd, Ollie Morton, Williard Brothers, Garvin E. Green, Elton Lowery, Johnnie W. King, Alfred Davenport, Donald Rice, Allen C. Dover, Hoyt L. Berry, Herbert W. Gargus

In the summer of 1961 when I worked at Tait's Gap, most of the men named above were employed, with the exception of William Carroll and Donald Rice, both of whom I knew at Woodstock; they had transferred to Tait's from Adkins Mine. I learned in late 2010 that William Carroll had passed away. He was my schoolmate at Greeley Elementary School in Caffee Junction. Also in 2010 we lost a member of the family and fellow miner Fred Cornelius, of Tait's Gap. I attended his funeral at Tait's Gap Baptist Church.

Tait's Gap Mine Shipments
June 1958 - Jan. 1959

	Cars of Ore	Cars of Tailings
June 1958	100	0
July 1958	100	6
August 1958	100	2
September 1958	91	2
October 1958	92	6
November 1958	99	3
December 1958	76	3
January 1959	86	1
TOTALS	744	23

Monthly Average = 93 cars/month of ore; 3 cars/month of tailings

Champion Mine Reopens, 1961

When mining operations resumed at Champion after 14 years, ore reserves previously unrecoverable were stripped of overburden using dragline excavators (Lima, 2-cu yd.) and hauled by the 22-ton capacity dump trucks (Euclid) to the cleaning plant, where the ore and rock were separated from the clay and sand matrix in log-washer vessels and multi-deck vibrating screens equipped with high pressure water sprays. The rock refuse was separated from the ore product by floating the lower density rock in a heavy-media suspension in which the higher density ore sank in the vessel of liquid heavy-media. This process resulted in a very precise separation of the ore and reject material, with only a minimal amount of misplaced product. The heavy-media consisted of a slurry of ferrosilicon (a magnetic iron powder), mixed in suspension with water to a specific gravity of approximately 2.9, in which the less dense rock floated and the heavier ore, having a specific gravity of 3.7, sank in the vessel, thereby effecting a separation of the reject from the ore. The rock and ore, on their separate paths, were rinsed off, the ferrosilicon recovered by a magnetic drum and recycled into the slurry mix of heavy-media. The clay-silt laden wash-water was pumped from the cleaning plant into the previously excavated ore pits and other sediment impoundments constructed for the clarification and recycling of the water. Water was in abundance at the Champion mine and no production was lost due to shortages of cleaning water for the plants, whereas dry weather had historically caused washer stoppages at nearby Tait's Gap.

My Summer Jobs at Tait's Gap and Champion - 1961-1962

In June 1961, I began my summer job at Shook & Fletcher's Tait's Gap Mine while between terms at the University of Alabama. My father, Howard C. Gunter, had been transferred there as assistant superintendent from the recently closed Adkins Mine at Caffee Junction in Tuscaloosa County, where I grew up and attended public schools. At Tait's Gap, we lived in a farm house belonging to my grandfather Ed Vandegrift. Mr. Vandegrift had been superintendent of the mines there since 1920, when Tom Worthington & Co. operated the mines for Sloss Iron & Steel Co. Shook & Fletcher Supply Co. of Birmingham took over the mines from Worthington in about 1921.

Tait's Gap was finally worked out after 41 years of off-and-on operations as dictated by the Birmingham steel economy. Shook & Fletcher put its hopes in reviving the old Champion property, last mined in 1944. They were still using 1920s-era steam shovels to excavate the ore although dump trucks had replaced the narrow-gage steam locomotives for haulage. Since their beginnings, the company's mining strategy was to go back into the relatively shallow old works with updated technology. In 1961, Shook & Fletcher leased the old Champion Mines property, jointly owned by Ed Vandegrift and Howard Gunter, who purchased it after WW II from Tennessee Coal, Iron & Railroad Co. (T.C.I.) and Sloss Iron & Steel.

A thorough drilling program had proven sufficient reserves so mining was begun and the ore hauled to the Tait's Gap washer for processing to evaluate the cleaning characteristics. The muck was excavated by dragline and hauled about five miles in off-road Euclid dump trucks along a gravel road along the L&N railroad for a testing period of about four months. As expected, it was quickly realized that the Tait's washer was not up to the task of separating the Champion ore imbedded in a sticky matrix of clay, unlike the more cherty and sandy material surrounding the Tait's ore.

After the muck was thoroughly wet in the dump box by a high pressure nozzle, the clay would become plastic and bind over the vibrating screens rather than separating from the coarse ore and rock material.

The oversize material not passing through the top screen, was directed onto a reject belt to a storage bin, intended to collect mud balls and limbs, stumps, and oversized rocks and ore not reduced by the crusher. This sticky clay containing the iron ore, however, would not break apart and pass the top screen as did the Tait's Gap muck, which contained much less clay. The washer, as designed and operating, could not handle the high percentage of oversized material and was rejected and directed to the mud ball bin. From this bin the mud balls were hauled by truck to recycle in the dump-box or dumped on a waste pile.

Champion Mine Conveyer Belt, man standing -picking out mud balls and rocks. right of heater..

I was assigned to the primary washer plant, under foreman Mr. Jim Dickie, a veteran miner approaching retirement. My job was to grease daily each of the dozens of bearings in the plant and empty the mud-ball bin as necessary. The bin seemed to fill every 30 minutes, and then I would load a single-axle five-ton Ford dump truck and recycle the mud to the dump-box. Soon I was overwhelmed trying to keep the bin from overflowing and the undersized Ford was replaced with a giant 22-ton Euclid dump truck, brought up from the Adkins mine. I was thrilled to be operating this awesome vehicle powered by a Cummins diesel, with automatic transmission, power steering, and air brakes, and horn. At last I could keep ahead of the mud ball flow into the bin, but the situation was grim for cleaning the ore at Tait's when, after hauling the ore muck three miles, you then dump about 15% of it on the waste pile.

Cleaning characteristics learned from this test period were applied in the design of the new Champion washer by adding more capacity for handling waste material. These same Champion pits had been mined off and on since 1888; therefore it was known that there were "leaner pickings" this time around and the new technology in excavators and ore preparation would have to make it profitable.

With the decision made to move operations to Champion, all production from Tait's Gap ceased and the washer dismantled with usable components salvaged for the new Champion washer. Other surplus machinery was moved there from Shook & Fletcher mines in Russellville and Caffee Junction. No more Euclid driving for me; now the real work began digging footings in hard chert to pour foundations for the new cleaning plant at Champion. Having mined on the Champion property since 1920, my grandfather knew exactly where the new plant should be situated for efficient wastewater clarification, clear water recycling, and sufficient elevation above the rail cars. A mile rail spur was built off the main line with a passing track for storing 25 cars. The only outside paid consultant hired to put in that mine, that I can recall, was a retired L&N Railroad track man hired to supervise the Champion miners installing the rail spur to the rigid specifications of L&N Railroad.

I still marvel at the skills of those miners, who got off their machines in July 1961, and constructed the new plant and support buildings with no outside help. Without any objections to engaging in strenuous manual labor rather than their routine machine operation, the miners rolled up their sleeves in the summer sun and went about constructing their new workplace. They did all the carpentry, concrete, plumbing, electrical, welding, and fabrication. All concrete was mixed on site using a one-bag mixer, using ore tailings for sand and gravel by-products. The hardest work I ever did was shoveling sand and gravel into that mixer for days on end, after digging the footings and building the forms. Lumber was sawed by Jesse Henderson on my grandfather Vandegrift's sawmill using oak, poplar, and pine timber cut from his farm. The primary washer structure used post and beam construction of white oak to support the heavy machinery. The work area and electrical controls were enclosed for weather protection.

Over the span of his career, Ed Vandegrift probably built a dozen or more wooden timbered ore washers, and this one may have been the last one built in Alabama. The life of a brown ore washer was only about five to ten years due to the nature of small ore deposits. Wood was available and low cost, and the small brown ore mines operated on low profit margins. In the late 1940s, a Shook & Fletcher mine in Russellville installed the heavy-media process to separate the ore from the rock, and in the mid-1950s this process was added at Tait's Gap to achieve precise separation of ore of ore rock. I recall the small drums in which the ferrosilicon powder was shipped. They were about 25-30 gallons capacity, weighed 500 pounds, and had to be lifted with the half-track boom truck of which I was the primary driver. Actually, the rear tracks on that boom truck had been worn out and the ever resourceful mechanics had welded a Ford dump-truck axle directly to the frame, minus any springs. Talk about a rough ride on those washboard roads!

In the late 1940s the Army surplus half-track personnel carrier was the best all-terrain "go anywhere" vehicle available for mining and lumbering, with the front wheel and rear track drive. At the Adkins mine, S&F bought about five of them for use in the drilling crews and as boom trucks utilizing their heavy duty front winch. One was even equipped with steel rail wheels and used as a locomotive at the bottom of a deep pit to shuttle a tramcar several hundred yards to the incline where the loaded tramcar transferred to a hoist cable for ascending to the top house to dump. While one loaded car was hoisted to dump, another empty car was towed by the half-track to the shovel for loading. That was the only open pit mine I ever saw that utilized a slope hoist and they mined in that one pit for 3-4 years, before having to branch out and go to Euclid haulage. I regret never having driven a half-track truck in original condition, but I did see them in action at Adkins hauling water to drill-rigs over muddy trails and woods. Even now, for off-roading, I'd choose a half-track over a new Humvee!! !

The heavy media process gave new life to the reserves there since the remaining ore was present with excessive rock which had to be separated before shipment to the furnaces. This process is also what gave Champion a final chapter in its long history. Even so it faced more difficulty in the summers, as did Champion prior to the 1960s, when adequate water impoundments were constructed by channeling a portion of Champion Creek flowing off Straight Mountain.

The heavy media plant was a pre-manufactured all-steel structure which was moved from Tait's Gap and set in place by a dragline on properly located foundation pads. It was used for primary cleaning that removes the residual sticky clay from the ore dirt that adheres to the equipment and slows the process. Refinement of the product makes it profitable.

Tait's had always struggled with insufficient water in the system, etc. I left the job in late August to return to the University, with the forty miners in full swing, constructing the new washer and infrastructure for getting back to mining ore, including sediment ponds, pumping stations, roads, water reservoirs, shops, office, rail spur, water system, etc.

Second Summer

When I returned in June 1962, for my second summer at Champion, my old washer foreman Jim Dickie had retired and Mr. Curt Dover was now washer foreman. Curt was a skilled carpenter who had built several barns and houses for my grandfather and now, after building all the frame structures at Champion, had settled down to running the primary washer plant. The heavy media plant was run by Lonnie Cornelius and Arthur Madden, both top-notch miners and individuals who ran their plant with utmost efficiency which led to the mines success. The ore analysis laboratory occupied one end of the office storehouse building; it was moved there from the Adkins mine along with chemist Ray Galbreath, a purple heart medal recipient in the Korean War. Ray's father was Webster Galbreath, who was the mine machinist/electrician/plumber and a favorite of mine. The timekeeper and parts manager was J. P. Bynum, Jr., a native who returned here after 20 years at the Doc Ray and Adkins mines. The pit foreman was Oscar Fullenwider, who began working at Champion in 1920, whose son Travis was a lifetime friend of mine.

My duties that summer were icing down the water coolers and delivering them to their stations, driving the fuel truck to the bulldozers, draglines, and shovels, and driving the road sprinkler truck on the Euclid haul roads. ,It was amazing how quickly those big "Euc's" could kick-up the dust again after a good thundershower., I had to refill the water truck about every 30 minutes from a pond using a temperamental hand-cranked four-inch Jaeger pump. Finally, I got the hang of how to start it and then I kept the dust down better. It now resides at my mining relics museum, along with the Jaeger cement mixer, 1930s Osgood dragline, WWII half-track, and 1948 F-5 Ford dump truck.

Another project I worked on in 1962 was installing a fine ore storage bin above the ore load-out conveyor so that the minus-3/8-inch ore from the jigs could be loaded in separate cars from the coarse ore. The bin held approximately 100 tons and was charged by a conveyor from the pair of Remer jigs, which separated the fine ore from the rock. The jigs produced about 100 tons daily of the minus-3/8-inch ore that the heavy-media process could not separate. Previously the furnaces had allowed the fine and coarse ore to be mixed together. Now the furnace companies were preparing briquettes of fine ore which would prevent the loss of fine ore forced upward and out of the furnace by the hot air blast through the bed of iron ore, coke, and limestone. On a "good" day the heavy-media could produce about five cars and the jigs about two cars, for typical 70-ton capacity railcars, amounting to about 500 tons/day. The primary washer could not consistently maintain that production rate, even by running 20 hours per day. I would estimate the average monthly mine production from 1961-1968 to be about 6,000-8,000 tons.

The cleaned ore quality was as good as ever, 50-55% iron, but more extraneous material had to be hauled, crushed, screened, and separated, just to get a ton of clean ore ready to ship. The cleaned coarse material sized 4" x 3/8" was separated in the heavy media plant where the ore sank and the lighter rock floated in a bath of ferrosilicon adjusted to a specific gravity of about 2.9. From this process, only about 1/3 of the product was iron ore. The cleaned by-product rock was sold locally for construction projects, and the tailings from the dried sediment ponds were sold to several cement companies as ingredients to manufacture cement.

In the spring of 1963, my parents moved into their new home built on the property by Curt Dover; it still stands on the pine tree-covered knoll in solitude among the flower gardens tended by my Mother. Two of the 60s era shop buildings are restored, as is the old office/supply/lab building.

According to Alabama Power Company, during the 1960s the Champion Mine was the second highest electric load in the county, behind the Inland Lake pumping station on the 60-inch pipeline to Birmingham Water Works. The large motors for crushers, pumps, screens, conveyors, log washers, and lighting were the major electrical equipment at the mine. The primary washer ran two ten-hour shifts daily for five days, with Saturday reserved for welding maintenance on the crusher teeth and worn metal surfaces.

Radio Hill 1961-1962

The foremen's cars from Adkins Mine were equipped with two-way radios, but the base station was still at Adkins mine in Caffee Junction where J. P. Bynum and Oscar Fullenwider were still dismantling that operation. H. C. Gunter found that he could drive his 1955 Ford to the highest elevation near the washer and communicate with the base unit at Adkins - a distance of about 60 miles. This high point (1,090 ft. above MSL –Mean Sea Level) was named Radio Hill and remains undisturbed since the 1920s, when it was last mined by steam shovels and Dinkey lines. The radio base unit was moved to Champion in 1963 and used with the three mobile units with great success.

Accidents

In October 1964, the main mud dam collapsed after several days of heavy rains and the outflow of muddy water and silt ran off in two directions, which was unusual. One path followed a northerly direction across a series of dams down the same hollow, taking out the two upper dams, with the two lower dams holding secure. When the torrent reached Hwy 132 culvert of Champion Creek at the Clint Payne (a S&F miner) residence, the water rose several feet and floated the floor furnace up into the house, awakening the horrified family in the predawn hours. They escaped to higher ground in their nightclothes by wading through the rising flood of mud as the house lifted off its foundation.

The other path overflowed a ridge and traveled westerly down Alabama Avenue in the now Eastwood subdivision, following the branch to the old Moody farm where it overflowed the highway and deposited a thick layer of silt and mud on the Tolbert farm and the old Moody airport field. By the grace of God, - no serious injuries were inflicted to residents, and miners set about relief work for the victims. Property damage suits were settled with Shook & Fletcher Supply Co. during the next few years. The cause of the dam failure was likely due to a partially blocked overflow pipe resulting in water breaching the dam following the deluge of rainfall for several days. Heroic efforts by Oscar "Pete" Hathcock to save the dam were too little-too late. Pete was the night shift dozer operator, pushing muck into the dump box, who upon realizing impending disaster at about 2:00 a.m., took his International TD-24 dozer to the overflowing dam, about 1/4 mile from the washer, and attempted to raise the dam by pushing material from the side. By then the dam was totally saturated and beginning to collapse, so fortunately, Pete saw his efforts were futile and pulled back before the dam was breached, sending torrents of mud down the valleys.

With all the dams and the pumping station washed away, mine management promptly set about to repair the damage and divert the muddy water to several deep worked-out ore pits as a short term solution. The long-term solution was to reactivate the large dry sediment pond on which my grandfather was growing record yields of corn to feed his herd of Polled Herefords. Levees were raised around the 100-acre cornfield and the 10-inch pipeline laid from the washer. Once again a previous mud pond from the 1920s was in use for the duration of mining. After raising the levees around the old dry sediment pond used in the 1920s, washer wastewater was then pumped to this impoundment where it was clarified and reused for the duration of mining.

Today this old sediment pond has about 2-3 feet of clear water above the estimated 30 feet of deposited washer tailings, covering an area of about 60 acres of lake and wetlands.

Euclid Overturned

Euclid Dump Truck

In an ore pit east (now a lake) of the L&N Railroad and within view of Highway 231, the Euclid dump truck of Coy Nobley overturned into the pit when it backed off a bank while positioning for loading by the dragline. He was pinned in the cab and mine rescuers Howard Gunter and Arthur Fullenwider first considered cutting Nobley free using an acetylene torch; however, that idea was rejected when they saw diesel fuel leaking. Dragline operator Howard Middlebrooks then moved his machine into position and lifted the truck upright with the bucket, thereby freeing Mr. Nobley, who escaped with only a dislocated shoulder. Sometime later at the same pit Mr. Nobley's Euclid narrowly escaped collision with the L&N mainline freight train after he failed to yield at the crossing, perhaps not hearing the locomotive's horn or misjudging the speed of the train.

Ore production averaged approximately 7,000 tons/month for the years 1961-1968, with additional leases with James Harp, City of Oneonta, plus old cuts along Champion Road.

Champion Mines Close for the Last Time

The last mining effort at Champion followed the closure of the Shook & Fletcher operations at Caffee Junction (Adkins Mine) and Tait's Gap. Many of those miners transferred to the new operation, including Howard C. Gunter, who with Ed Vandegrift had purchased the mine property from T.C.I. and Sloss-Sheffield after WW II. The reopened Champion Mine featured large diesel-powered draglines, bulldozers, and Euclid dump trucks to strip, load, and haul the ore muck to an improved ore washer utilizing a "heavy media process" previously proven at Russellville and Tait's Gap mines.

By comparison to the boom times of the late 1920s, final mining efforts in the 1960s produced rates of only about 6,000 - 8,000 tons/month, due to deeper excavations and leaner yields from the "muck" hauled from pits. However, this rate was maintained for seven years until the last ore train pulled out of the Champion rail spur in 1968, to end an 80-year lifespan of the mine. The ore shipped in the final year, 1968, was 55,000 tons. The quality of the cleaned ore never faltered from that which Col. Henry DeBardeleben once named it, *Champion*. Although production records are missing for several years, it is estimated that the quantity mined from the Champion District, 1889 to 1968, was 2,500,000 tons.

The final chapter closed with the advent of competition from the low-cost, high-grade Venezuelan iron ore (70% iron) imported through Mobile to Birmingham. Since 1968, nature has. reforested and filled with wildlife the barren knobs and dried sediment ponds, and filled the many lakes and ore pits with clear water. The L&N trains no longer roar through Murphree's Valley carrying coal, limestone, and iron ore for the steel industry, but the memories of those of us who experienced some part of the fascinating era of Champion Mines will remain forever.

Miners with whom I (Van Gunter) Worked - Summers - 1961 & 1962

Fred Cornelius
Hobert Henderson, dragline operator
Alton Henderson, Euclid driver
Jim Dickie, washer foreman
Clarence Cornelius, mechanic
Jesse Fullenwider, dozer operator
Gordon Hutchins, dragline operator
Howard Middlebrooks, dragline operator
Clarence Davenport, Euclid driver
Alfred Davenport, Euclid driver
Earnest Clements, bulldozer
James Tidwell, washer crew
Pete Hathcock, bulldozer

Grady Hathcock
Verlon Tipton, Euclid driver
Coy Nobley, Euclid driver
Dewey Clements, mechanic
Herbert Gargus, dozer operator
Verben Morton, shovel operator
Oscar Walker, shovel/dozer operator
Clint Payne, washer operator
Jack Clements, track crew
Curt Dover, carpenter/washer foreman
H. Gargus, dragline operator
Lonnie Cornelius, heavy media plant
Arthur Madden, heavy media plant
Lossie Byrd, washer crew

Glossary of Ore Mining Terms

BENEFICATION – concentrating the ore by removing unwanted waste materials by mechanical means; i.e., washing.

DINKEY - a small narrow-gage steam locomotive used to transport ore from pit to washer. The type used at Champion was a 0-4-0 drive arrangement, 36 in. gage, with saddle water tank and coal box aboard.

DORNICK - (locally pronounced "donick") a boulder, especially one of iron ore found in limonite deposits.

JIGS - apparatus in ore-cleaning plant which separates fine particle sizes of ore (minus 3/8") from waste rock by passing a slurry of fins across a vibrating table thus causing a gravity separation.

HEAVY MEDIA - ore cleaning process utilizing a suspension of water and finely ground magnetic iron oxide powder (ferrosilicon) adjusted to a specific gravity of about 2.9, in which the lighter rock floats in a rotating cylindrical vessel of heavy media, while the heavier ore sinks to the bottom. The magnetic powder is recycled by washing off both the "float" and "sink" products and recovering it using a rotating drum magnet.

LIMONITE (brown ore) is an oxide of iron; sometimes called brown hematite (Hematite-red ore is the most common form of iron ore). The iron content of the Champion ore ran consistently in the 55% range, whereas, hematite, the predominant ore mined in the Birmingham District typically contains 40% iron. Also, the desirable high content of manganese (1.3%) with low content of undesirable phosphorus (0.25%) and silica kept this ore in high demand for the pig iron market.

LOG WASHER – ore-cleaning device consisting of long, narrow, inclined vessel having parallel shafts counter-rotating and having intermeshing paddles.

MSL – Mean Sea Level

MUCK - Material loaded in ore pits to be transported to the washing/benefication plant

MUCK BOX - permanent structure constructed at head end of washing plant where dump trucks or trams unload muck to be cleaned, open-topped, three sided inclined box lined with steel plate, which converges to narrow discharge into crusher. The plant feed-rate is regulated by an operator using a gimble-mounted, high-pressure and high-capacity water nozzle, which dissolves muck, causing it to slide by gravity into the crusher.

MUCK POND - impoundment for containment of waste-water from ore washing process, where clay and fine ore and rock particles (minus 3/8") are settled from suspension and the resulting clarified water is recycled to the cleaning process. Mud pumps were used with cast iron pipe lines to transport wastewater to the sediment ponds and abandoned pits.

MUCK-RAKER - laborer stationed on muck pond dam, raking muck from pond to top of dam to dry and thereby build up height of dam above water level for containment of muck. Also, a laborer stationed at "picking belt" conveyor at discharge end of revolving conical screen in an ore washer, who rakes off mud balls and rocks ahead of the crusher.

PIG IRON - iron that has been run directly in "pigs" or sand molds from a blast furnace; cast iron.

STEAM SHOVEL – rail-mounted or crawler-tread-mounted mechanical crane with dipper at end of shipper shaft. (Early Champion machines were rail mounted; S&F era used crawler-treads.)

TAILINGS – the fine material which is discharged with the muddy water during the washing process and stored in sediment ponds.

References

1. Historic Sketch and Incidents in the Life of D. B. Bailey, self-published, Altoona, Alabama
2. The Story of Coal and Iron in Alabama, Ethyl Armes, 1904
3. Some Recollections of Champion Mining Camp, by Aulden Woodard, 1996
4. Diary of Ed Vandegrift, Mine Supt., Tait's Gap and Champion Mines, 1928-1965.
5. Powell's History of Blount County, 1855, George Powell
6. Recollections and articles - Iron Age, Mary Gordon Duffee, Blount Springs, ca. 1900.
7. A Profile of Gabriel Hanby, Blount County Historical Society, 1962
8. Unpublished documents, by Edward V. Gunter
9. Tannehill and Growth of Alabama Iron Industry, J. R. Bennett, 1999
10. Alfred Montgomery Shook Family History, by Anne Kendrick Walker

Section Six

MEMORABILIA – ARTICLES - PHOTOGRAPHS
FROM THE MUSEUM'S ARTIFACTS FILES

The Champion Mines

Map - Blount County

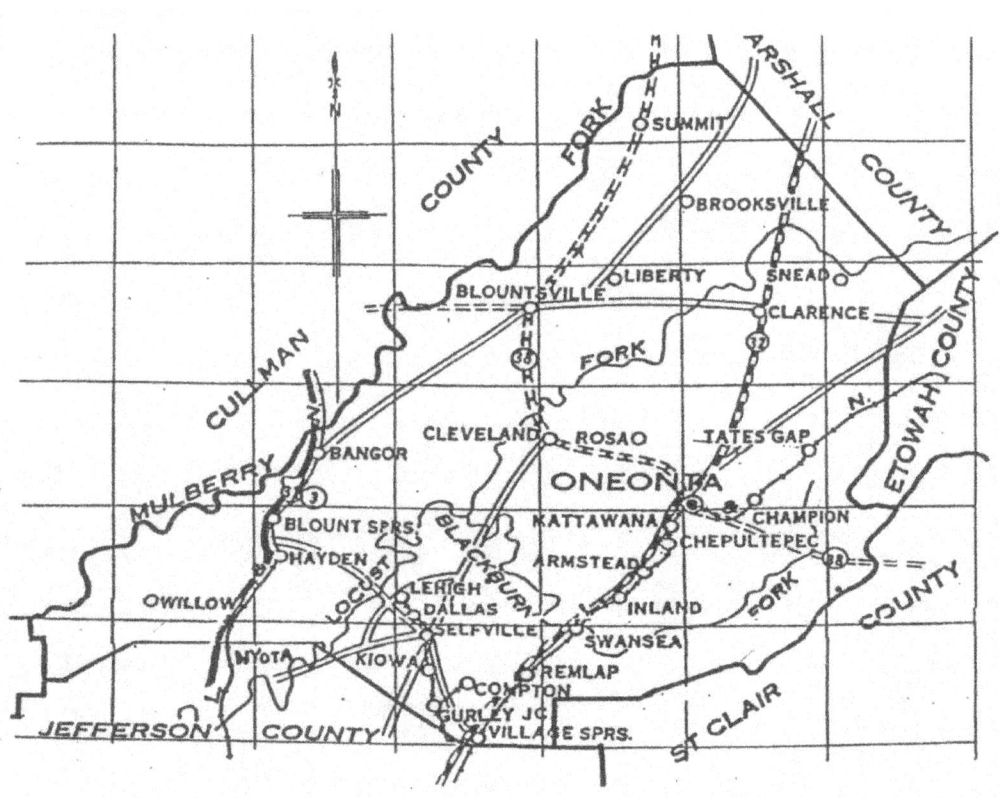

The Champion Mines

CLIPPING:. THE BLOUNT COUNTY IRON INDUSTRY?

Former Champion miners Alfred Davenport, Garvin Green, Howard Roberts, Fred Cornelius, H. Wayne Gargus, Hobert Henderson, and Verlon Tipton. (Not shown is Alton Henderson)

The Blount County iron industry?

by Wallace Todd

Most Blount Countians are not aware that an area here was once the site of a booming iron mining community. East of Oneonta and north of what is now U.S. 231 south lay the expansive Champion Mine. In fact, the existence of a mining community there predated the incorporation of Oneonta by some two or three years.

Van Gunter, historian and current owner of the property where the mine was located, is a curator of photos, documents, and artifacts from those days of yore. He is organizing the Champion Miners Society and has begun a mining museum at the site. As a result of his extensive research of company records and consultation with former employees of the company, he has compiled what he terms "The Champion Hall of Fame."

The list, in chronological order of ownership, begins with John and Gabriel Hanby. It is believed that the two brothers were descendants of a prominent Virginia family who came to the area in the late 1700s. According to Gunter's chronology, the Hanby brothers discovered rich, high-grade iron ore in the area in 1817. The first recorded owner of the site was James D. Crump.

In the early years of the mine's history, no efficient mode was available to haul the ore to processing facilities in Birmingham. It is believed that local blacksmith shops rendered the iron and molded weapons, horseshoes, cannonballs, and other implements for General Jackson's troops during the Civil War.

Hobert Henderson, one of the veteran miners, owns an antiquated shingle sawmill that was apparently marketed by the Oneonta Machine Works. He theorized it was made from Champion iron, cast in the Birmingham area, then assembled in Oneonta in the early 1900s.

By 1882, Henry DeBardeleben and James Sloss bought the land and brought to the area a spur line of the L & N Railroad. As a result of the mining industry, the town of Oneonta had sprung up and, after heated debate and two public votes, the county seat was moved from Blountsville to Oneonta.

The mine changed ownership several times between 1891 and its closing in 1968. Nat Baxter and Alfred Shook bought it in 1891 and from

See IRON INDUSTRY, pg A12

The Champion Mines

Iron Industry clipping (continued)

Oneonta City Council.
2008 tax levy passed
The commission approved the 2008 annual tax levy at a rate of 26 mills per $100 of taxable property for citizens living in the county. Nine mills of the total goes to the general fund to cover expenses of the county, and two mills goes to the Blount County Health Care Authority to support public health projects and facilities. The remainder goes toward the support of city and county schools. A 10 cent tax per acre on forestland is

reason for the action was to eliminate the redundancy of two separate annual inspections.
Routine items
In other actions, the commission:
• awarded the bid for chip-spreader work and operator to A-1 Operations with a bid of $155 per hour;
• awarded the bid for Emergency Management Agency interior and exterior surveillance cameras and associated electronic equipment for the courthouse to Forge Technologies with a bid o_

Core samples from Champion mine

IRON INDUSTRY
CONTINUED FROM FRONT PAGE

1925 until 1945 James and Paschal Shook owned it. Edgar Vandegrift, Van's grandfather, and Howard Gunter, Van's father, were its owners until 1968.

Gunter divides the mine's history into four distinct eras:
Manual, mules and oxen – 1817 to 1889
Steam and rail – 1889 to 1923
Electric and internal combustion engines – 1925 to 1944
Diesel and heavy media – 1961 to 1968

The mine's peak output years were between 1925 and 1967 when it was operated by Shook & Fletcher under the direction of E. N. Vandegrift. Gunter said that at one time the mine employed around 400 workers. During the mine's most productive years, the company complex amounted to a small independent community with its own school, stores, and around 44 residences where most of its employees lived.
Employee reunion
Recently, a group of veteran miners met at the Blount County Memorial Museum in Oneonta to reminisce about their hardships, antics, and camaraderie at the mine. Hobert Henderson said, "They were all good people. We were like a family. We stuck together and helped each other when they needed help."

Steamshovel and locomotive

brought some grins and silent chuckles. One said he started out at $13.20 for a 60-hour week. By 1947, some were knocking down as much as $28 for a 60 hour week. All agreed the work was hard and dirty, but the friendships that were established there overshadowed adversities.

Gunter's developing Champion Miners Society is searching for others who may have documents, photographs or any other connection to the mine. He has a display in the Blount County Museum containing scores of ore samples and other items connected to the iron ore industry. Amy Rhudy, curator of the museum, suggests the display would be an ideal focal point for science field trips, as well as history and geology enthusiasts.

More information about the history of Blount County mining is available by calling Gunter at (205) 625-4098 or Rhudy at (205) 625-6905 or by visiting the museum, open Tuesday through Fri

The Champion Mines

CHAMPION MINERS SOCIETY REGISTRATION FORM 2004 – 2010

***** REGISTRATION FORM FOR MEMBERSHIP *****

CHAMPION MINERS SOCIETY
Dedicated to the Memory of Blount County Ore Miners, 1882-1868
Est. July 25, 2004
Historic Champion Mines ~ 110 Gunter Lane, Oneonta, AL 35121

Please indicate you wish to join the *Champion Miners Society* by checking this space: _____

Name(s): _____

Address: _____ City/State/ZIP: _____

Phone: _____ E-mail (if any): _____

Family members (or Friends) who were Champion or Taits Gap Ore Miners (if any):
_____

Please indicate if you have old newspaper articles, photographs, documents of Champion Mines, which you would share with this organization ---
_____

Please list other family members you wish to include as members of Champion Miners Society. There are currently no membership requirements or dues for this organization:
_____

-------------------- detach here --------------------

*** Complete form & mail to: VAN GUNTER, 110 Gunter Lane, Oneonta AL 35121 *****
(please copy this form and distribute to others as needed)

CHAMPION MINERS SOCIETY
"Dedicated to the memory of Blount County Ore Miners, 1882-1968"
CHARTER MEMBERS ~ 1ST Gathering, July 25, 2004, Champion Mine Office

Fred Cornelius, ex-miner ~ & wife	Alfred Davenport, ex-miner
Wayne Gargus, ex-miner	Garvin Green, ex-miner
Hobart Henderson, ex-miner	Van Gunter, ex-miner ~ & wife Sara
Verlon Tipton, ex-miner ~ son & 2 gndchldn present	

Wayne Shaddix ~ son of Miner Cecil Shaddix, gndson of Miner N. I. Shaddix
Tim Chamblee ~ Oneonta -- WKLD
Lawrence & Jane Gunter Doughty ~ daughter of Miner Howard C. Gunter
*Jesse & Mary Nell Holt ~ Oneonta, BCHS
*Terry & Sue Fullenwider ~ son of Miner Arthur Fullenwider
*Walton V. & Helen Linder ~ g-son of Miner E.N. Vandegrift
*Mayor Ralph Tidwell & wife Sue ~ son of Miner Arthur Tidwell-Champion Camp
*Comm. R. C. Smith, III, Blount Co. Commission

The Champion Mines

Clipping: L&N HELPED START ONEONTA IN 1888

L. & N. helped start Oneonta in 1888

By A.A. FENDLEY, Mayor
and W.J. WOODS, L&N Agent
(An Account From the 1940s)

By 1900 the town had become a thriving trading point and somewhat a cotton market. In 1903 the first bank was established. The year 1908 saw the establishment of the Blount County Bank, now a leading institution of the community, having in 1926 taken over the oldest bank of the town. In 1914 the Farmers Savings Bank was established which in 1921 became the First National Bank. The Oneonta Guaranty Company, a semi-banking institution, was started January 1, 1926. The town now boasts of three strong financial institutions with over a million dollars on deposit.

The Brown Ore Mine at Champion and Tait's Gap, the Cheney Lime Plant and Youngstown Mining Company, all in the immediate vicinity, not only add thousands of dollars to the business of the town, but show the close proximity of the materials that have made Birmingham a great iron and steel city. Two large wood working plants in town add to the pay rolls and increase the business and wealth of the community.

The Hotel Garner, one of the finest small hotels in any small town in the state, was built early in 1927, and surprised its owners by showing a profit almost from the day it opened.

Almost $100,000.00 has been spent in new school buildings and churches in the last two or three years, and over thirty new modern residences are now under construction.

The City Council has just passed an ordinance providing concrete paving over the entire business district and part of the residence section. State Highway Number 44 to Birmingham, thirty-six miles, was completed two years ago, and survey is now complete to extend the road to Boaz which will connect for through travel between Birmingham and Nashville.

The L.&N. favored the town by giving us a through train to Birmingham and Gadsden on one hour schedule each way. Merchants have sold more goods this fall than ever before and are more prosperous.

The people here believe it is one of the best towns in the Birmingham district and an excellent place to live.

The L.&N. started all this in 1888 when it built its Mineral branch through here from Birmingham to Gadsden and Anniston.   +

HOTEL GARNER, ONEONTA, ALA.

ALA—5 HISTORICAL CHRONICLES OF SOUTH, Week of December 1, 1975

Chapter XIII

Growth in Mining Areas

AN IMPORTANT nucleus for the L. & N.'s further growth had been created in July 1884, following the completion of the Birmingham Mineral Railroad. The latter (owned and operated by the L. & N.) had a modest beginning and, as its name would imply, was projected to serve the iron and coal industries of the Birmingham District. It was originally only 11 miles long and consisted of two branches, the North and South. However, the Company had ambitious plans for this seedling and it was the intention to encircle Red Mountain, thereby tapping the rich mineral deposits and serving the many industries which had sprung up mushroom-like in the vicinity. The original trackage extended from a junction with the S. & N. A. at Magella, Ala., (about three miles south of Birmingham) for a distance of about seven and one-half miles along the northern base of Red Mountain to what later became the town of Bessemer, so named with prophetic insight by H. F. DeBardeleben, its founder. The South Branch led off from the S. & N. A. about four miles south of Birmingham at Graces, Ala., and skirted the southern base of Red Mountain for a distance of about three and one-half miles to Redding, Ala. As mentioned, these two small lines (later to be linked together by additional construction) served as bases for the L. & N.'s many and extended beneficial forays into the Birmingham District.

By the year 1886, the city of Birmingham, a lusty adolescent of 15 years, was experiencing phenomenal growth. Nearby towns, too, like Bessemer (1887) sprang up overnight and for a brief space seemed to threaten the very supremacy of Birmingham itself, but in the end they but contributed to the greater glory of the Magic City. Speculation was rife and one would have to go to the Florida boom of the 1920s for adequate comparison. But whereas the Florida boom was largely the result of real estate speculation alone, Birmingham's 1886-1887 crashing crescendo was a hectic mixture of the selling of real estate, the mining of coal and the making of iron, with other less important items flavoring the heady brew of growth and expansion.

The Magic City's glittering future, as well as the demands of the traffic, seemed to justify the L. & N. in its construction of a new depot there and this was completed in 1887, at a total cost of $134,163.95, this sum representing expenditures not only for the station itself, but for passenger tracks, a trainshed, etc.

The Railroad's faith in the future of Birmingham and the Birmingham District was thus based not alone on wishful thinking. Its Birmingham Mineral Railroad, which received its initial impetus from Milton H. Smith, was rapidly bringing it in touch with many flourishing iron and coal operations and the Annual Report for the fiscal year ending with June 30, 1887,

84

The Champion Mines

GROWTH IN MINING AREAS - ARTICLE XIII (LOUISVILLE AND NASHVILLE RAILROAD: 1850 - 1963 BY KINCAID A. HERR - 1964 CONTINUED

states that at that time there were in existence upon the lines of the L. & N. and those of its sister road, the N. C. & St. L., a total of 32 furnaces producing huge quantities of pig iron, 11 of these furnaces, utilizing charcoal and 21 of them using coke. In the Birmingham District alone there were 33 coal and iron companies. There were also under construction at the time 28 additional furnaces, only six of which were of the charcoal variety. It was estimated that the coke furnaces then operating could each produce 115 tons of pig iron a day, while the charcoal ones had a rated daily capacity of 50 tons each. As a matter of possible interest, the recipe for one ton of pig iron at that time was two tons of iron ore, one and one-half tons of coke, and one-half ton of limestone.

Prominent among the industries at that time (1887) in the stronghold of iron and coal were the Tennessee Coal Iron and Railroad Company, which was just completing four large furnaces at Ensley known as the "Big Four," the Pratt Coal and Iron Company (soon to be absorbed by T. C. I. & R. R. Co.), the Sloss Furnace Company, the DeBardeleben Coal and Iron Company (capitalized at $13,000,000), the Cahaba Coal Mining Company, the Pioneer Mining and Manufacturing Company (later to become a part of Republic Iron and Steel), the Woodward Iron Company and the Eureka Furnace Company, each representing an investment of many hundreds of thousands or, in some cases, millions of dollars.

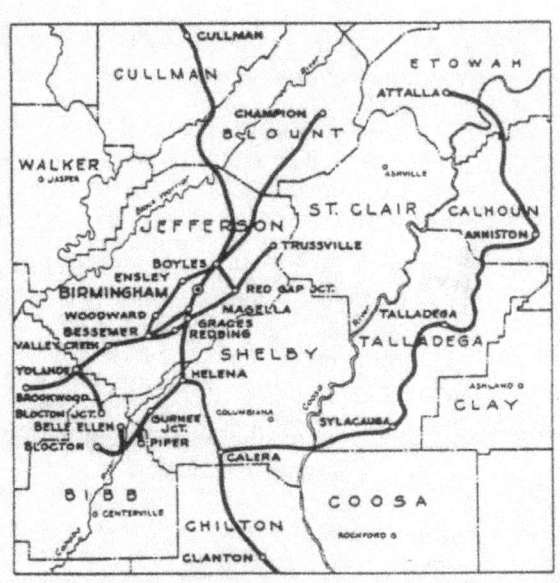

The L. & N. lines in the land of coal and iron in 1890.

These properties were owned or operated by men who have since become almost legendary figures in the world of coal and iron: Daniel Pratt, Truman H. Aldrich, Henry F. DeBardeleben, Enoch Ensley, James H. Sloss, William T. Underwood (the brother of that Senator, Oscar W. Underwood, for whom years later Alabama Democrats were so persistently to cast their 24 votes for Presidential nominee), John T. Milner, T. T. Hillman and many others.

While it is true that the manufacture of pig iron and the mining of coal monopolized the picture, there were a number of other flourishing industries in the Birmingham District and their number was constantly being increased. For the most part these industries manufactured products such as pipe, car wheels, axles, stoves, nails and hardware of all sorts and descriptions for whose making large quantities of pig iron were required. Numerous rolling mills were also established to serve as middlemen between the furnaces and the factories.

In this connection it is interesting to note that steel was first produced in Alabama as early as 1888. On March 8, of that year, the Henderson Steel and Manufacturing Company produced a ton of that metal, utilizing ordinary Alabama iron ores. This steel was subsequently successfully used in the manufacture of razors, carving knives, etc., but the fruits of this achievement were highly deciduous. In all, the Henderson Company produced about 1,800 tons of fine steel, but the cost of operation was entirely out of proportion to the price that could be obtained for the product in the open market and in 1890 the company's furnace was turned over to a committee from the Birmingham Chamber of Commerce as a possible proving ground for the introduction of steel-making into the Birmingham District. One of this committee was Pulaski Leeds, the L. & N.'s superintendent of machinery. This committee reported favorably upon the furnace and its process, but outside capital strangely enough was somewhat coy and the plant was subsequently abandoned. It was years later, in 1897, to be exact, before steel was manufactured again in the Birmingham District. But, of that, more later.

Other railroads, too, were being attracted to Birmingham, each anxious to serve as Mercury to Vulcan. Lines which later became parts of the Southern, the Central of Georgia and the Frisco were all completed to Birmingham during the 'eighties. Thus, the position of the South and North Alabama Railroad, long the dominant one of the district, was being challenged by these new arrivals.

A less alert leadership might have allowed this early-gained and hard-won advantage to be forfeited through sheer inertia or a too-fond recollection of past glories. But that was not the policy of the L. & N. Its rapid-fire construction of innumerable branch lines and industrial spurs more than kept it in the running. Nourished by L. & N. capital, the Birmingham Mineral Railroad, that rapidly-growing satellite of the S. & N. A., shot out tendrils of steel and wood in every direction, allowing the Iron Horse access to remote mountain fastnesses where large coal mining operations were being prosecuted or where huge quantities of red and brown hematite ore were being removed from the bosom of Mother Earth.

This account, endeavoring as it does to preserve a judicious balance of power between bare statistics and the more colorful aspects of our Road's building, will not attempt to list in iron-clad detail the gradual evolution of the Birmingham Mineral Railroad. This has already been done by very competent authorities. Then, too, some of these branches, spurs, etc., once projected to serve then vital needs, have long since been abandoned and nowadays no trace of them remains. Others have been sold to other railroads. In a broad way, however, we shall endeavor to sketch the growth of the Birmingham Mineral Railroad as it occurred year by year in that territory which largely lies to the southwest of Birmingham.

Mention has been made heretofore of that "basic" trackage which skirted the south and north bases of Red Mountain, just south of Birmingham. These two branches were soon connected, forming a loop of some $18\frac{1}{2}$ miles around Red Mountain. (A portion of this trackage, from Redding to

The Champion Mines

GROWTH IN MINING AREAS - ARTICLE XIII (LOUISVILLE AND NASHVILLE RAILROAD: 1850 - 1963 BY KINCAID A. HERR - 1964
Continued

An L. & N. flat car proudly carried this 11-ton lump of Alabama coal to the World's Industrial and Cotton Centennial Exposition at New Orleans in 1884. It was shipped by the Pratt Coal and Iron Co., near Birmingham. Reading from the left, the men are: Col. Enoch Ensley, L. W. Johns, Joshua Collins and William Gude.

Bessemer, was later abandoned.) On January 1, 1888, the North Branch which extended to Bessemer was further extended to Blocton Junction, some 27 miles distant, by way of Valley Creek and Yolande, this being known as the Blue Creek Extension. In addition to the trackage previously mentioned, construction was completed of a line from Bessemer to Boyles, (Huntsville Branch No. 1) nearly 16 miles long and extending through the thriving communities of Woodward and Ensley. A line (east of the S. & N. A.) from Boyles to Red Gap, a distance of about 6½ miles, was also built.

Then, in 1889, or thereabouts, some 60 additional miles were added to the Birmingham Mineral's trackage, making in all a total of 132.60 miles. Included in such construction was a line from Boyles to Champion, a distance of 36 miles, and the further extension of the road from Red Gap to Trussville, some 11 miles away. During the fiscal year ending with June 30, 1890, some 24 additional miles were constructed, resulting in the Birmingham Mineral then having a total trackage of 156.22 miles to its credit.

Various spur lines were constructed in conjunction with these branches and the Birmingham Mineral and the South and North Alabama between

87

The Champion Mines

GROWTH IN MINING AREAS - ARTICLE XIII (LOUISVILLE AND NASHVILLE RAILROAD: 1850 - 1963 BY KINCAID A. HERR – 1964 CONTINUED

them rendered a real service to the industries of the District. Then in 1890, they were joined by a potent ally, the Alabama Mineral Railroad Company, which was incorporated on July 28, 1890, with the L. & N. owning a majority of the capital stock. This venture was the result of the consolidation of the Anniston and Atlantic Railroad (incorporated on May 24, 1883) which boasted of some 53 miles of narrow gauge track, extending from Anniston to Sylacauga and the Anniston and Cincinnati Railroad (incorporated on January 31, 1887) which ran between Anniston and Attalla, a distance of 35 miles. Both of these roads were built by A. L. Tyler and Samuel Noble, the founders of Anniston, that city (Annie's Town) being so named in honor of Mr. Tyler's wife.

Each of these roads was acquired by the L. & N. shortly after completion, or on July 19, 1889, and it immediately set about to correlate their activities with the rest of the System. The formation of the Alabama Mineral Railroad was the first step towards this end. The Anniston and Atlantic's narrow gauge line was changed to four feet, nine inches, the 30-pound rail was replaced with 58-pound rail and the line was changed at a number of points to reduce grades, eliminate curves, etc. New construction work was also commenced in March 1890, the line eventually being extended from Sylacauga to a connection with the South and North Alabama at Calera, 34 miles south of Birmingham. This linking was completed on January 1, 1891, with the result that the Alabama Mineral Railroad now consisted of 119 miles of main line track and 13.23 miles of spurs and branches, extending from Calera through Sylacauga, Talladega and Anniston to Attalla. These lines formed, when viewed on the map, a rough-hewn masculine profile which, with some pardonable imaginative license, suggests the features of the "Father of our Country."

This left a gap of some 26 miles between Attalla and Champion, which at that time was the northern outpost of the Birmingham Mineral's Huntsville Branch No. 2. The L. & N., for obvious reasons, was extremely anxious to complete its "great circle," embracing St. Clair county in totality and large portions of Shelby, Talladega, Jefferson and Calhoun counties. The Alabama Mineral Railroad served a territory which was rich in marble, limestone, brown hematite ore and other mineral deposits and to move these commodities to Birmingham or to the South and North Alabama from Attalla, Gadsden, Anniston and other points on the "forehead" of the profile was a roundabout procedure. There were other railroads providing short cuts to Birmingham from various points on the Alabama Mineral; nevertheless it was May 28, 1905 before the missing link of the Alabama Mineral was completed between Attalla and Champion, providing a more direct route to Birmingham via the L. & N.'s Alabama Mineral.

Much iron ore and coal also moved over the Alabama Mineral Railroad from Birmingham and points nearby to the numerous iron furnaces in operation at Attalla, Gadsden, Anniston, Talladega and Shelby. It was also planned that the cars moving onto the Alabama Mineral after being unloaded, could be re-loaded with the brown hematite ores heretofore mentioned for shipping to the Birmingham District for subsequent admixture with the red hematite ores, it having been found that

Clipping: Blounty Couty's minerals and water furthered Birmingham's industrial development

UNTIAN / SECTION B / ONEONTA, ALABAMA / DECEMBER 31, 1957

t County's minerals and water furtherec
ngham's industrial development

ames like "Champion" and "Nyota" and
back to the county's mining history. In
es to the mining industry, like "Bangor

Oneonta, got its name from "the
cording to an article by David Brewer.
f Blount County in Birmingham's iron
nning during the New South era follow-

earth in the Champion mining opera-
cond only to Russellville ores. Some
e Russellville ore, for reasons that did
's research. By those accounts, therefore
on."
in the area bearing that name today,
either gave or took from the area its

stone of high purity that lay in a
ath into Shades Valley, in Jefferson
ble for its quality and its accessibility

antial mineral deposits in Blount Coun-
ngham's industrial boom following the
d growth in the 20th century.

Birmingham industrial engine with all
id Brewer. (See related article about
t.) Thus Blount's growth and prosperity
mingham's progress.
Brewer's article — interesting reading
beginnings, development, and role

Coal came out of a Lehigh mine by the carload in 1924. Lehigh was a top producer from Blount County to the Birmingham district steel industry.

Another source, Ethel Armes in her *The Story of Coal and Iron in* ... Lehigh mines produced ...

The Champion Mines

Clipping: Blounty Couty's minerals and water furthered Birmingham's industrial development

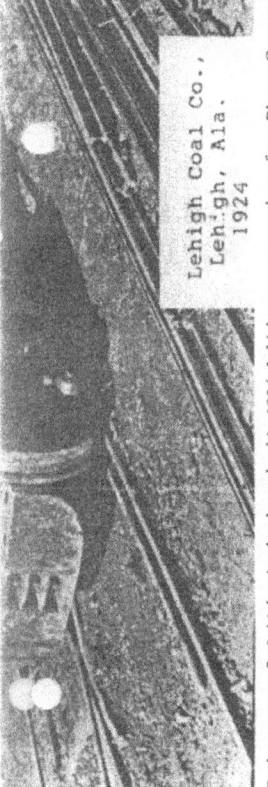

Champion, Al. 1928
Transportation was primitive in 1928, but the job got done.

Lehigh Coal Co., Lehigh, Ala. 1924
Coal came out of a Lehigh mine by the carload in 1924. Lehigh was a top producer from Blount County to the

site ... from ...th of Bangor south into Shades Valley, in Jefferson Cou... the ...estone was notable for its quality and its accessibility along... ...als.

Those ...ther areas of substantial mineral deposits in Blount County played a c... H...al role in Birmingham's industrial boom following the Civil W... we... B... continued growth in the 20th century.

AL FOUR...er

In fact, Blo... nt alone fueled the Birmingham industrial engine with all ...out ... ...ource ...s, according to David Brewer. (See related article about Brewer, a Blount County resident.) Thus Blount's growth and prosperity became closely coupled with Birmingham's progress.

Following are highlights from Brewer's article — interesting reading about Blount County's industrial beginnings, development, and role today in the region.

Tens of thousands of tons of minerals were dug from Blount County and sent to Birmingham (and later to Gadsden's steel plants). In the latter decades of the 19th century, the boom began with a 12-fold increase in pig iron production in Birmingham between 1880 and 1890. In that era Birmingham boasted 29 of Alabama's 53 blast furnaces. In 1892 those furnaces produced 994,000 tons of pig iron, a crude iron cast in blocks.

Blount's resources had become evident earlier, according to Brewer's article. An 1855 account by George Powell in *A Description and History of Blount County* cites "some very fine beds of ore in Murphree's valley," luckily near "limestone, good fire stone, and a good coal bed."

Another source, Ethel Armes in her *The Story of Coal and Iron in Alabama*, says John Hanby, who served with General Andrew Jackson in 1812, discovered the brown ore near Oneonta.

RAILROAD THE KEY

But the resources weren't deeply utilized until a railroad brought transport. The Birmingham Mineral Railroad laid the first line 36 miles out of Jefferson County to Champion in 1889. The L&N bought the BMR in 1890, added 12 miles beyond Champion, and offered reduced rates for shipment of ores.

With railroad service, the big movers in Birmingham started buying up mineral rights in the county. Notable names include DeBardeleben, the Sloss Company, the Tennessee Coal, Iron, and Railroad Company (TCI), and the Alabama Steel and Wire Company of Birmingham. Holdings were sometimes in thousands of acres.

FIRST MINE

The first mine "of any extent" is recorded during 1892-94 as Swansea, six miles south of Oneonta, with veins of "good coking coal." The Champion operation first produced brown iron ore in 1888. The Compton red ore mine in southern Blount County produced in 1883, but dropped in production in 1894.

Limestone began to be quarried near Bangor and Blount Springs by 1892. Limestone is important to the refining process as a "flux" or flowing agent, which combines with impurities in the ores and flows these out, according to Brewer.

1900 AND ON

Operations continued to expand after the turn of the century. Tonnage increased several fold. For example the Lehigh mines produced about 36 tons in 1904, and 92,000 tons in More coal and iron companies tion" in county records.

Limestone operations expanded. Cheney brothers from Anniston. They purchased limestone outcroppings south of Oneonta and convinced the L&N that it should build a spur track, shipping the fi... barrel in 1903.

In part due to this support from Blount County, Alabama won first place in the nation for pig iron p... duction, following Pennsylvania, Ohio, Illinois, and New York.

MEN AND MACHINES

Thus far, these minerals had be... extracted from the earth through strenuous, backbreaking, manual labor put forth by hundreds of workers. Only occasionally do ac... *The Southern Democrat* of 1917,

In 1911 the first machines cam... efficient, machines no doubt con... the following years. The Nyota m... in production. In 1912 these dril... between 1915 and 1920, annual in 1920 the mines bore almost 1...

In 1911 Blount's limestone pro... only to Shelby and Etowah coun...

The Champion Mines

Miscelaneous Clippings

Oneonta Southern Democrat

December 13, 1928

A VALUABLE INDUSTRY

While there are many industries shut down in the Birmingham district, and others running half time and one third time, there is one in the Oneonta district that is not only running on full time, but has for more than a year been running day and night in order to fill the demand for its products.

We refer to the Shook & Fletcher Supply Co. operations at Champion and Tait's Gap.

This concern under the management of Mr. E. N. Vandergrift, has expanded from a small beginning a few years ago until now it employs about three hundred men and has a monthly payroll of $30,000.00.

From these mines comes some of the highest grade brown ore to be found in any part of the country. It is a grade of ore that there is always a market for. When the mines were first opened in 1888, the ore produced was regarded by mining engineers and chemists as the best ore of its kind in the world. From this the place derived its name—Champion—the champion ore of the world.

After being in operation for a number of years, for some cause, work was suspended for a time. When electric power became available through the Alabama Power Co., the mines were re-opened by Mr. Vandergrift and both the Champion and Tait's Gap plants electrically equipped.

This industry means much to Oneonta and the surrounding country. Its annual payrolls equals one fourth the entire cotton crop of Blount county.

While Mr. Vandergrift does not employ many high priced men they are good honest citizens who meet their obligations and men who are never in the courts for violations of the law.

Another thing that makes this a valuable industry for Blount County is the fact that nothing used in the development or operation of these mines has been bought outside of the county that could be bought in the county.

An industry like this, though small, is a distinct asset to the county. We need more of them and we need more men like Mr. Vandergrift.

May 24, 1917

New Ore Mines Being Opened

The opening of a new ore mine about one mile north of Oneonta is to begin this week.

The Worthington Construction Co., with about 50 mules and 50 men arrived here Wednesday to begin the work.

The opening is to be on the Armstrong tract on Red Mountain. This is a fine grade of ore and has been mined on a small scale at different times for a number of years, but owing to the fact that it is off of the railroad nearly a mile it has not been found very profitable.

It is understood that if the opening is satisfactory a railroad will be built to the mines.

August 13, 1925

W. E. CALVERT HURT IN MINES

W. E. Calvert, whose home is on route 1, Trafford, was seriously hurt in the mines at Nyota last Friday when he was caught beneath a falling rock. It is said that his back was broken. He was carried to the Norwood Hospital, Birmingham, where attendants say there is a chance for his recovery, but his injuries are very serious.

Mr. Calvert is 29 years of age and is a son of Rev. J. J. Calvert. He has a wife and one child.

March 12, 1925

WORK BEGINS AT CHAMPION

Work was started Monday on re-opening Champion mines which has been idle for the past two years. They will be operated by Shook & Fletcher Supply Co., one of the biggest mine operators in the Birmingham district.

E. N. Vandergrift will be superintendent, handling the project in connection with the Tait's Gap operations. The plant will be electric lighted and operated by electric power. About 75 hands are to be employed.

The mines at Champion are said to produce the highest grade ore in the world, the name "Champion" was given it because of this fact.

November 12, 1925

POWER LINE TO BE EXTENDED TO ALTOONA

The Alabama Power Co., will ask the Public Service Commission for permission to extend its transmission lines from Champion to Altoona and to establish a schedule of rates for serving the people of Altoona.

Champion mines are now being operated by power furnished by the Alabama Power Co. and Cheney Lime Co. will use the power as soon as motors can be installed.

The Champion Mines

CLIPPINGS (CONTINUED)

Oneonta Southern Democrat

March 26, 1936

WORK STARTS ON BIG DAM

More Than Thousand Men Now Employed Making Excavation

Work started during the past week on the big dam at Inland which will impound the waters of the Blackburn Fork of Black Warrior river for Birmingham's Industrial water supply.

More than a thousand men are now employed making excavations for the dam. They are taken from the relief rolls and most of them are from Jefferson County. A special train of some fifteen or twenty passenger coaches is used to transport them to and from their work.

The dam, which will be located about eight miles south of Oneonta, will be one of the biggest projects of its kind in the State. The length of the dam will be 600 feet and its height 180 feet. The thickness at the base will be 120 feet; at the top it will be 20 feet. It is estimated that 195,000 cubic yards of cement will be used in the structure.

To reach the site of the dam it will be necessary to build two miles of roadway. This will leave the Birmingham - Oneonta highway near the Hawkins place and will go up the river on the South side. The road is to be properly graded and drained and hard surfaced.

The water from this dam will be conveyed through sixty inch steel pipes for 16 miles to a point near Mt. Pinson to a distribution reservoir. The water will flow by gravity to this reservoir.

The distribution reservoir will hold about 130,000,000 gallons, or two days supply at capacity consumption. From this reservoir, the water will be carried by gravity to industrial Birmingham. The line will enter the industrial district at Tarrant City and will be carried through Birmingham and westward as far as Woodard.

The reservoir behind the impounding dam at Inland will hold between 22,000,000,000 and 25,000,000,000 gallons of water. The lake will run back into the Straight Mountain section a distance of seven miles. Some of the back waters will come within four or five miles of Oneonta.

The system will have a capacity of 60,000,000 gallons daily.

May 20, 1937

Inland Dam

This story on construction work at Inland Dam appeared recently in the Birmingham News.

Construction crews are now working 24 hours a day in an effort to complete the impounding dam for Birmingham's $7,000,000 industrial water supply system by Christmas.

Contractors, however, have until March, 1938, to complete the work City Engineer J. D. Webb said. Construction of the pipe lines, part of which are being built by contract and part with WPA forces will be completed well in advance of that time, he said.

A concrete lined water division tunnel, through which the river will flow while the dam is being constructed is now nearing completion. This tunnel will be 20 feet in diameter and part of it will be used for the spillway when the dam is finished.

The dam located about a mile from Inland in Blount County, will be of rock filled clay construction. It will be 200 feet high and 950 feet thick from the water face to the back at the bottom and 30 feet thick at the top. Across the stream it will be 1,100 feet long. It will contain 1,000,000 cubic rods of rock and 600,000 cubic yards of clay.

Water backed up by this dam will cover more than 1,600 acres. It will back water 7 1-2 miles up the river channel. The lake will be two and one-half miles across at its widest point.

When completed the system will furnish 60,000,000 gallons of water a day to industries of the Birmingham district. The lake at Inland will store 21,000,000,000 gallons of water. A storage reservoir near Ketona will have an additional capacity of 60,000,000 gallons.

A 60-inch pipe line will extend from the impounding reservoir to Twenty-seventh Avenue and Vanderbilt Road. Smaller lines will then distribute water over the industrial area and the line will continue to Woodward, where it ends.

Mr. Webb said 80 per cent of the portion of pipe lines being constructed with WPA forces already are complete and the remainder are scheduled to be finished by the end of June. About a quarter of a mile of 60-inch pipe line being constructed by contract remains to be finished.

As soon as the water division tunnel is finished construction of the dam will be rushed.

Work is in charge of the Engineering Commission composed of City Engineer J. D. Webb, A. Clinton Decker and O. O. Thurlow.

February 19, 1970

50 YEARS AGO...

New Mine Being Opened At Taits Gap. Work on the opening of a new mine at Taits Gap was commenced last week by Tom Worthington & Co. About thirty hands are now at work and the force will be increased as the work progresses until more than a hundred men will be on the pay-roll. Modern machinery including steam shovels and washers will be installed, involving an expenditure of approximately $100,000.00. The new mine when in full operation will employ a large number of men and it is said that from fifteen to twenty cars of ore will be shipped dailey.

The erection of a number of houses to take care of the employees will begin at once.

The ore being mined is the same grade of that now being mined at Champion.

The new ore mine is on the west side of the railroad while the big D. B. Gore coal mine is on the east side.

The Champion Mines

Post Cards

POST CARD

BLAST HOLE DRILL.

This machine is at work at the top of the quarry face of the Cheney Lime Company. Holes 6 inches in diameter and 125 feet deep are drilled at intervals of 20 feet. Several hundred pounds of dynamite are charged into each hole and from 1 to 10 holes are shot in a battery. Each hole should produce 5,000 tons of stone, which must be re-drilled, broken up again with dynamite and finally hand broken and selected for the manufacture of Cheney Lime.

Much of this stone is crushed and used for furnace and foundry flux and for concrete and road building.

CHENEY LIME COMPANY,
Allgood, Ala.

Post Cards of Quality.— The Albertype Co., Brooklyn, N. Y.

THIS SPACE FOR MESSAGE.   THIS SPACE FOR ADDRESS.

The Champion Mines

Postcards (continued)

of brown i... I found a rich seam 1882 when Henry DeBardeleben and James Sloss sought land and brought L&N Railroad causing county seat to be moved from Blountsville to Oneonta in 1889. Most ore was mined by Shook and Fletcher 1925–1967 from Champion + Taits Gap mines under E.N. Vandegrift, superintendent. Ore was shipped to Woodward. T.C.I + Sloss furnaces in Birmingham and Republic in Gadsd[en]

The Champion Mines

Mines Ad - Tait's Gap Coal

TAITS GAP COAL
NOW READY FOR DELIVERY

Lump Coal 6 Inches and up
Egg Coal, 1 1-4 to 6 Inches
Mine Run Coal, all sizes
Double Screen Stoker 5-16 to 1 1-4 in.
Double Screen Stoker 5-16 to 1 1-4 in. Oil Treated

Buy Your Coal Now To Protect Yourself Against The Rush and Possible Shortage!

Certified Weights and Satisfaction Guaranteed

MAIL YOUR ORDER TO:
ALTOONA, ALABAMA, RT. 1

J. W. PHILLIPS, Salesman

Robbins Coal Co.

The Champion Mines

Map   Champion Map Camps – 1935 - In 3 page parts.

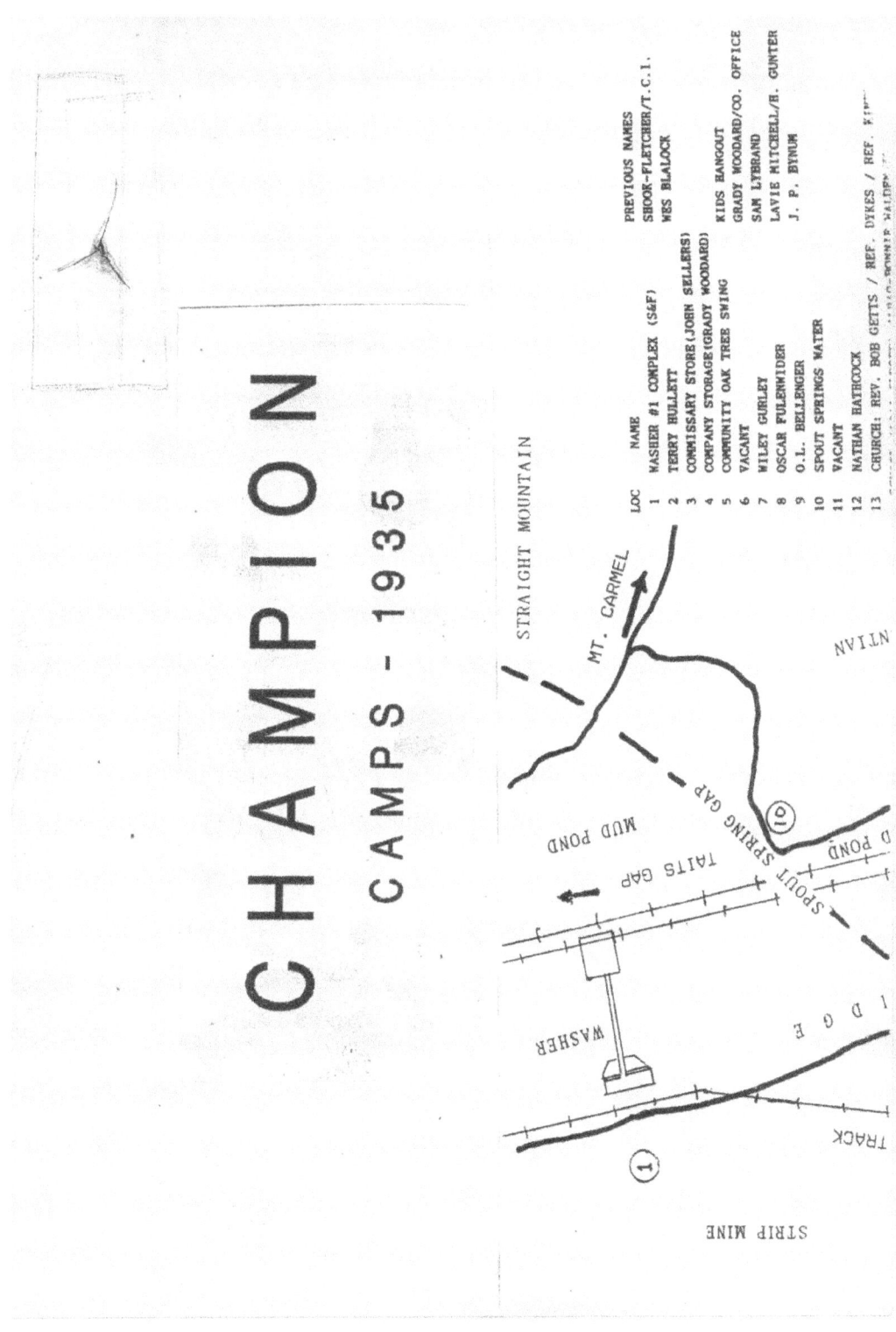

The Champion Mines

Campion Map. (continued)

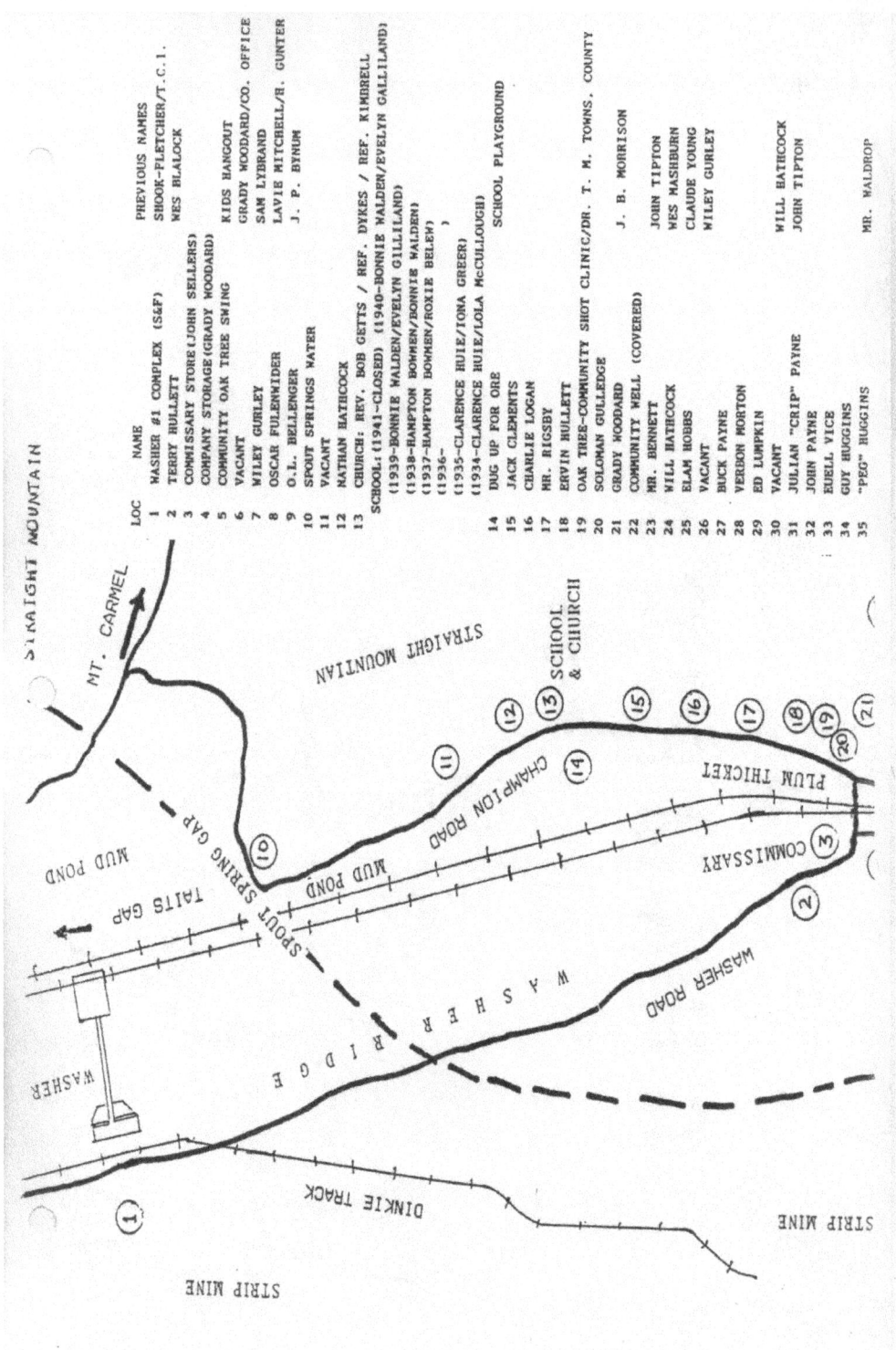

The Champion Mines

Champion Map (continued)

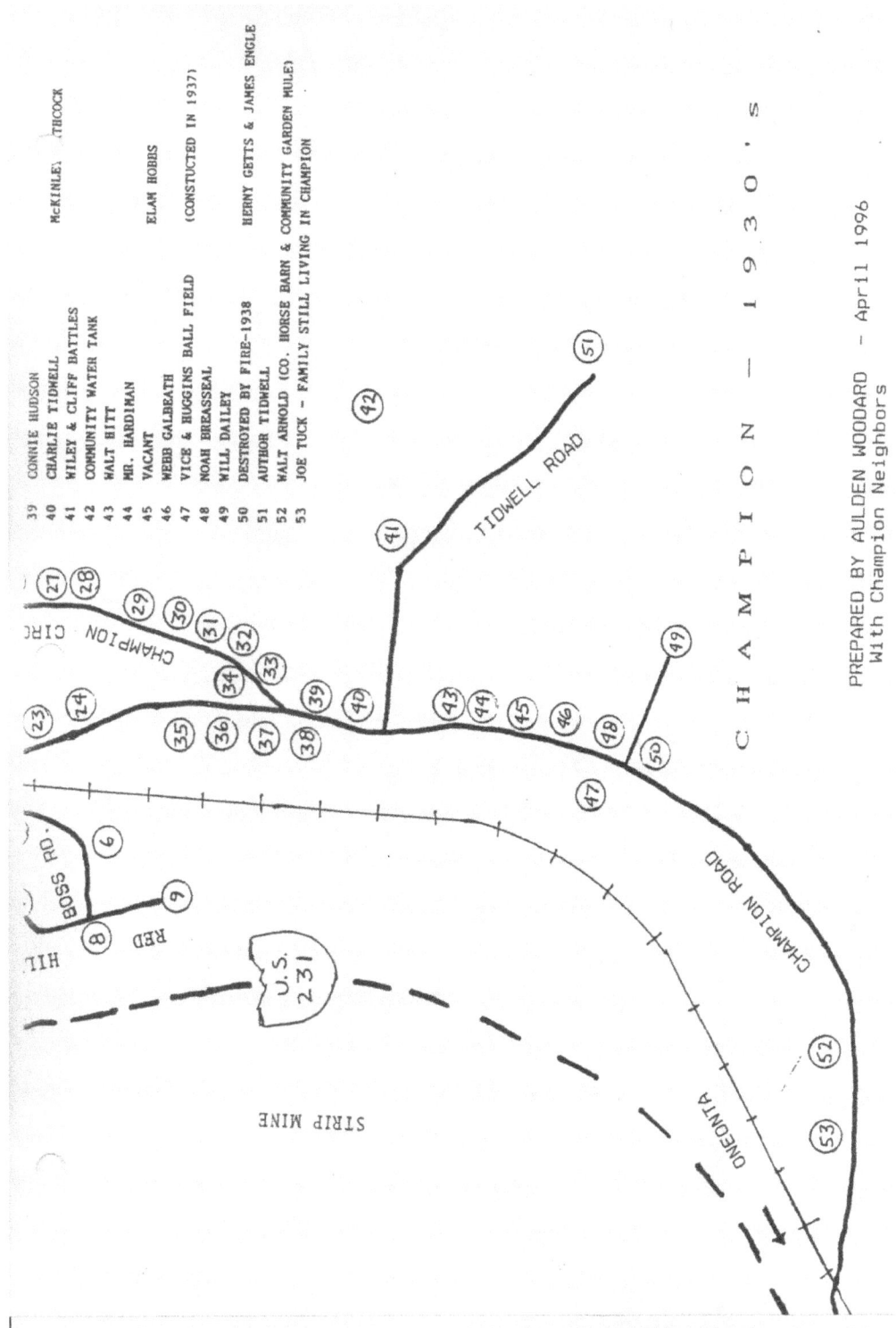

The Champion Mines

DRAFT - LETTER FROM DAVID T. PALMER TO DR. PAYNE, 7 NOVEMBER, 1996. BLOUNT COUNTY: A GEOLOGICAL SURVEY-

David T. Palmer

Dr. Payne

GS 302

7 November 1996

                    Blount County: A Geological Survey

    Blount County is geologically varied and interesting. The soils are divided into five general categories. The county has a varied supply of mineral deposits and is sprinkled with large and intriguing caves. Anyone who enjoys exploring the earth, especially a geologist, would find a visit to Blount County a worthwhile experience.

    Being politically and geographically older than the state of Alabama, Blount County at first extended from the Cherokee line on the northeast to the present Tuscaloosa line on the southwest. It included Jefferson and Walker Counties and a large part of Marshall County (Blount 8). However, today Blount is bordered by these counties in addition to Cullman, Etowah, and St. Clair counties. Today, Blount County covers 640 square miles and lies in the Cumberland Plateau section of the Appalachian Plateau area of Alabama. The county is divided geologically into six districts. Going from northwest to southeast, these districts are the Warrior Basin, Sequatchie Valley, Sand Mountain, Murphree's Valley, and Blount Mountain (Heritage 7).

    The Locust Fork and Mulberry Fork of the Black Warrior River serve as the major part of the drainage system of the

The Champion Mines

Palmer Letter. (continued)

county. However, the north central part of the county is drained by Big Springs Creek and Brown's Creek (Heritage 8). These rivers and creeks also serve as sources of recreation to the people of Blount County.

Blount County's soils can be categorized into five general groupings based upon their origin. The first, limestone soils, range from Brown to Big Springs valleys and are noted for row crops and pastures. The second type, cherty limestone, can be found on rolling to hilly land and is good for growing farm crops or timber. Sandstone, the third type, constitutes some of the best soils of the county and is excellent for row crops and highly responsive to fertilizers. Shale is the fourth type and is found on gentle slopes and can be used for pasture, timber, and some row crops. The last soil, mixed origin, produces good pasture and grain crops (Heritage 8).

The value and extent of the mineral deposits in Blount County are not completely known. Red iron ore occurs in beds extending northeast and southeast in the eastern half of the county. The beds average more than two feet in thickness. Brown iron ore can be found in the southeastern part of the county near the red ore deposits (Heritage 8). During the young years of the county, Mr. John Hanby, a machinist from the Blount Springs area, discovered a pocket of brown ore in the neighborhood of Oneonta. "This rich find was years later acquired by Major Tom Peters, and sold by him to Henry F. De

The Champion Mines

Palmer Letter. (continued)

Bardeleben and Colonel Sloss, and named Champion" (Armes 23). In its prime, the mine yielded much ore.

Extensive deposits of coal are found on both sides of the Locust Fork River, extending northeast and southwest (Heritage 8). During the late 1960's and 1970's, extensive strip mining for coal was done in various sections of the county. In 1855 George Powell, one of the earliest settlers of Blount County, wrote, "The coal beds that I have seen are about two feet thick and of good quality" (Blount 8). Other minerals found in the county are limestone, dolomite, manganese, sandstone, chert, and Chattanooga shale (Heritage 9).

People who enjoy spelunking can spend hours on end exploring the different caves Blount County has to offer. Some of the caves were used by the Aborigines as burial places, and their remains can still be found in them along with fragments of lead, mills, shell, and trinkets (Heritage 10).

The Crump Burial Cave near Clear Springs was discovered in 1840 by James Newman. He was out hunting when he stumbled upon the cave. It is situated in the steep limestone cliffs where the Warrior River enters a gorge and leaves the valley. When Newman first entered the cave, it was about four hundred feet above the river and fifty feet below the plateau above. A short distance from the entrance stood a room, which turned out to be a burial site of the aborigines. Newman found eight or ten wooden coffins of

The Champion Mines

Palmer Letter (continued)

black and white walnut. The coffins were sent to the Smithsonion [*late 1800's by Frank Burns, Blountsville*]. Also found near the coffins were human skulls and bones (Heritage 10).

One of the more popular caves in Blount County is Bangor Cave. It was once used as a gambling and whiskey party spot. It is so large that there was room enough for a live band and a dance floor. People partied in the cave in the 1920's and 1930's until the local sheriff shut it down. Rumor has it that Clarke Gable once partied in Bangor Cave.

Another cave, Rickwood Caverns, is a state park. Tours of the cave are conducted by park employees. The cave is a popular tourist spot.

Blount County has many points of geological interest. Its caves are fascinating, and its cliffs are high and wide. There are farm lands and mines, forests and rivers. Blount County is truly beautiful and interesting.

Works Cited

Armes, Ethel. The Story of Coal and Iron in Alabama. Leeds, Alabama: Beechwood Books, 1987.

Blount County: Glimpses from the Past. The Junior Blount County Historical Society, 1989.

The Heritage of Blount County Alabama. Reunion Edition. Blount County Historical Society, 1989.

Geological Survey of Alabama
Brown Ore Production – Champion District
One ton = 2240 pounds

Year	Mine	Annual Tonnage	Avg. Price $/ton	State Production, tons
1910	Champion	67,624	1.53	1,123,136
1911	Champion	< 50,000	1.47	835,886
1912	Champion	<50,000	1.43	
1913	Champion	<50,000	1.61	
1914	Champion	<50,000	1.49	
1915	Champion	55,427	1.68	
1916	Champion	<50,000	1.97	
1917	Champion	51,906	3.08	
1918	Champion	<50,000	3.55	
1919	Champion		3.52	
1920	Champion		4.26	
1921	Champion idle	0	2.50	
1922			2.55	
1923	Tait's Gap		2.89	
1924	Tait's Gap		2.74	
1925	Tait's Gap & Champion			
1926	Tait's Gap & Champion	148,429	2.74	
1927	Tait's Gap & Champion	197,446	2.58	
1928	Tait's Gap & Champion	247,255		

The Champion Mines

BROWN ORE PRODUCTION BY YEAR

Alabama Department of Industrial Relations
Brown Ore Production by Year

Year	Mine	Employees	Tonnage	S&F Adkins Mine	S&F Doc Ray
1968	Champion	38	54,490		
1967	Champion	39	71,620		
1966	Champion	34	78,470		
1965	Champion	33	72,522		
1964	Champion		96,218		
1963	Champion		65,525		
1962	Champion		45,823		
1961	Tait's Gap		31,569		
1960	Tait's Gap				
1959	Tait's Gap		49,981	146,713	
1958	Tait's Gap		68,347	33,754	
1957	Tait's Gap		51,218	216,071	
1956	Tait's Gap		41,888	224,532	
1955	Tait's Gap		28,438	215,645	
1954	Tait's Gap		27,328	209,477	
1953	Tait's Gap		32,549	172,713	
1952	Tait's Gap	23	35,713	223,731.	
1951	Tait's Gap	23	37,704	167,524	
1950	Tait's Gap	22	39,202	252,748	
1949	Tait's Gap	21	25,000	200,000	
1948	Odenville	25	29,570	62,106	**Tonnage**
1947	Odenville	27	49,381		118,423
1946	Odenville	26	32,003		76,000
1945	Champion	25	15,828		89,071
1944	Champion	26	15,000		140,863
1943	Champion	25			
					102.868
Combined Champion & Tait's Gap			1,094,990		
Add Odenville			1.205.944	3.25 mill.M tons	0,53 mill. M tons

Vandegrift mined approximately 5 million tons of ore as superintendent during his S&F career, 1921-1965.

196

The Champion Mines

LETTER FROM SHOOK & FLETCHER

Fig. 6 - Letter from A. M. Shook III

Shook and Fletcher Supply Company Records, Box #1042.
Department of Archives and Manuscripts, Birmingham Public Library

SHOOK & FLETCHER SUPPLY CO.

MOLONEY ELECTRIC COMPANY
TRANSFORMERS
ALLIS CHALMERS MFG. CO.
MOTORS, PUMPS, TEX-ROPE DRIVES
OIL CIRCUIT BREAKERS
THE CENTRAL FOUNDRY CO.
UNIVERSAL CAST IRON PIPE AND FITTINGS
AMERICAN BRASS COMPANY
COPPER AND BRASS TUBES
BARS, RODS, SHEETS
CARRIER CORPORATION
REFRIGERATION, AIR CONDITIONING
MORGAN ENGINEERING COMPANY
CRANES & STEEL MILL EQUIPMENT

DELTA-STAR ELECTRIC COMPANY
DISCONNECT SWITCHES, BUS SUPPORTS
TERMINATORS
LIDGERWOOD MANUFACTURING CO.
HOISTING ENGINES, STEAM, ELECTRICAL
CONTRACTORS EQUIPMENT
MACKINTOSH-HEMPHILL CO.
ROLLS AND ROLLING MILL MACHINERY
PENNSYLVANIA ELEC. COIL CORP.
ELECTRIC COILS
S. P. KINNEY ENGINEERS
BRASSERY SELF CLEANING WATER STRAINERS

ORE MINING

1814 FIRST AVENUE NORTH
PHONE 3-6281   P.O. BOX 2631

THE OHIO BRASS CO.
RAIL BONDS, INSULATORS
ELECTRICAL TRANSMISSION SUPPLIES
LA-DEL CONVEYOR & MFG. CO.
MINE CONVEYORS, VENTILATING EQUIPMENT,
MECHANICAL LOADERS
M. W. KELLOGG CO.
HIGH PRESSURE PIPING AND VESSELS
ANACONDA WIRE & CABLE CO.
BARE AND INSULATED
WIRE AND CABLE
SHENANGO-PENN MOLD CO.
CENTRIFUGAL CASTINGS
BUCKEYE LABORATORIES CORP
INSULATING & LUBRICATING
OIL PURIFIERS

Birmingham, Ala.,
January 14, 1946

Dear Uncle Paschal:

I am enclosing a rough draft from the Tennessee Co. on proposed brown ore lease for the Woodstock district, and ask that you please look this over at your convenience and I would like to hear from you in regard to other changes than I have made that you think would be necessary for our own good. There is one thing that Blair failed to include in this which I have not put on the margin, and that is the fact that they have agreed to take all ore down to 40%. There is an awful lot of stuff in this thing that seems to me is unnecessary; however, if they insist on it and you feel that it is all right, we will of course go ahead with it.

As soon as we received word Saturday night that the steel strike had been postponed, we got hold of the mines and railroads to have cars in so that we could start running Monday morning. I am glad to report that all mines are running today. I was talking with some of the boys at T.C.I. this morning out at the coal mines and they all feel that there will be no work stoppage; that this thing will be settled or compromised in some way. I certainly hope that their assumptions are correct.

We sold one of the old steam shovels this morning for $1,000.00 cash, as is, where is, to the Robbins Coal Co. at Altoona, and already have their money and it is in the bank.

We also received a check for $3,029.75 from the Swindle Coal Co. at Lynn, Alabama, covering one shaker conveyor, paid in full. You will recall that you prepared notes for these people to pay over a period of time but they have decided to pay cash for it.

Adkins is leaving tomorrow for Knoxville, flying up, and expects to get back in here Thursday night or Friday morning, based on what reservations he can make. No other news here. We are all well, and my love to Aunt Carrie.

Love,

Mr. P. G. Shook
The Waves
Delray Beach, Florida

AMS/em

Fig. 6

The Champion Mines

ECHAMPION MINES HISTORECAL MARKER TO BE DEDICATED SUNDAY

This photo graph, dated 1929, shows a washer overlooking the L&N Railroad at old Champion Mines. Dinkey cars are pictured dumping ore to be washing.

A4/THE SOUTHERN DEMOCRAT/Oneonta, Alabama/July 26, 1989

This photograph, dated 1929, shows a washer overlooking the L&N Railroad at old Champion Mines. Dinky cars are pictured dumping ore to be washed.

CHAMPION MINES HISTORICAL MARKER TO BE DEDICATED SUNDAY

A Champion Mines historical marker will be dedicated at Blount County Historical Society's quarterly meeting Sunday, July 30, 2 p.m., at Blount County Memorial Museum here.

A century ago, the mineral line of the L&N Railroad was extended to the Champion Mines, making it possible to ship the large quantities of high-grade brown iron ore first discovered by John Hanby, who served under General Andrew Jackson in the War of 1812.

Between 1886 and 1890, there was a boom in the mineral belt. Along the railroads where the resources seemed superior, towns were laid off with streets and avenues. Oneonta was designed and laid off for a city, and the railroad, the minerals, soils, and stones gave birth to the City of Oneonta, which was incorporated in 1891. The city was named by William Newbold, L&N superintendent, because it reminded him of his hometown of Oneonta, N.Y.

The Champion Mines historical marker granted by Alabama Historical Association gives a brief history of the development of the industry, which began with John Hanby's discovery. It reads as follows:

"John Hanby came in 1817 and found a rich seam of brown iron ore. Named Champion in 1882 when Henry DeBardeleben and James Sloss bought land and brought L&N Railroad, causing county seat to be moved from Blountsville to Oneonta in 1889. Most ore was mined by Shook and Fletcher 1925-1967 from Champion and Taits Gap mines under E.N. Vandegrift, superintendent. Ore was shipped to Woodard, T.C.I., and Sloss furnaces in Birmingham and Republic in Gadsden."

John Hanby's brother Gabriel came to Blount County with him and was the county's first state senator.

"The History of Champion Mines," a paper prepared for Alabama Historical Association by Emma Linder in 1987 and presented at the annual meeting, was published in *The Southern Democrat* in January 1988 giving a detailed account with pictures. This can be seen at the museum.

The marker will be placed on U.S. 231 at its intersection with Champion Road across from the entrance to Eastwood subdivision. The city and its Neighborhood Council are preparing the site and plan some landscaping there.

All those who worked at the mines have a special invitation to attend the historical society's meeting Sunday.

The Champion Mines

Champion Historical Marker Unveiled

Champion Mines marker unveiled

The Champion Mines

Clipping Citizen Shook left Mark

Thursday, September 8, 1966 — The Birmingham News Daily Magazine

WALLING KEITH—
Citizen Shook left mark

Paschal Shook's life spanned several industrial revolutions. And in his way he helped contribute to some of them.

He came down from Tennessee with other members of his vigorous family in the early days of the Tennessee Coal Iron & Railroad Company and for years the fortunes of the Shook family, TCI and Birmingham were intertwined.

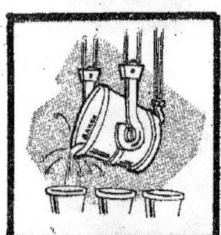

Paschal Shook's footmarks were embedded in the hard Birmingham soil when the city was small and struggling. The marvel of it is that he kept freshly apace with a metropolitan city and a world that so many of his contemporaries never lived to see.

He was a stirring figure in Birmingham in my boyhood days and knew the birth of Ensley and Fairfield and helped found numerous enterprises that have helped build this whole area. He came along in the days of the rugged individualist, knew and dealt with the powerful men who made steel a great and growing thing in America. And among his friends were the early ironmasters and coal barons.

His life closely touched men like Col. Enoch Ensley, G. B. McCormack, Erskine Ramsay, Victor Hanson, Oscar Wells, Morris Bush and others who helped make a strong city stronger.

Until a few months ago you could see him on the streets of the city he loved so well and helped so much, a tall, military-shouldered man looking 30 years younger than his age.

AT THE AGE OF 90 I USED to see Paschal Shook of a Saturday morning sometimes at the big downtown Birmingham postoffice. He liked to go by on Saturday mornings and pick up his mail, return to his office and work and plan.

On several such occasions I had fruitful chats with him about our city and state, about the background for some of the wonderful institutions we have learned to take for granted.

KEITH

He knew Birmingham's past well because he had been a part of it. But he dwelt keenly in the present and was interested in tomorrow.

There are not many of his day and his kind left in Birmingham. Or anywhere else for that matter. And those of us who try to understand today a bit better by the events of yesterday sorely

In troublous times I have

In troubleous times I have found it strengthening to talk with men like Paschal Shook who stood up against so many misfortunes and near-disasters that touched this community over the years.

Sometimes I get to thinking that men and communities are shaped and toughened together. It is difficult at times to tell which does the most for the other.

Anyway I feel all the better for knwing that there are strong places here and there in our town made all the stronger by men like Paschal Shook who stood straight and tall for his 94 years — and most of those years in Birmingham.

Paschal G. Shook, Chest founder, dies

Paschal G. Shook, 94, founder and president of Shook & Fletcher Supply Co., died at his home Thursday.

Graveside services will be held at 10:30 a.m. Saturday at Elmwood Cemetery, Johns Ridout's Southside directing. The Rev. Hugh Agricola of the Episcopal Church of the Advent will officiate.

Born in Nashville, Tenn., in 1872, Mr. Shook came to Birmingham in 1895. With John Fletcher, he founded the Shook & Fletcher Supply Co. in 1900. He was president of the company until his death.

ELECTED A director of First National Bank of Birmingham in 1915, Mr. Shook served in this position until 1965.

He was also a director of Alabama By-Products Corp. and served on its board until his death.

A founder of the Community Chest, Mr. Shook served as its president and for 23 years as **PASCHAL G. SHOOK** a member of the board of directors. He was elected an honorary life director of the Chest.

Surviving are two daughters, Mrs. Allen A. Johnson and Mrs. Charles L. Gaines; a son, Paschal G. Shook Jr.; seven grandchildren and three great-grandchildren, all of Birmingham.

The Champion Mines

Tait's Gap School – October 2011 – with new entry/porch

Mr. Dowd, Washer Foreman, and his car – Champion Mines 1929

The Champion Mines

FAVORITE PAST TIME – SITTING ATOP CHAMPION'S RAILROAD STATION SIGN

CHAMPION – PUMP POND 1928 – SITTING ON INTAKE HOUSING IN POND

The Champion Mines

Vehcle Service Checklist, 1968

Check When Serviced	Vehicle	_	Anti-Freeze added later								Anti-freeze required
___	Jethro's Lima	11-3-67 / 12 GAL	11-7-69 / 2 GAL	12-28 / 1 GAL	1-19-69 / 1 GAL	2-13 / 3 GAL					12 gal.
___	Buddy's Lima	10-28-67 / 12 GAL	11-8-69 / 1 GAL	12-15-69 / 2 GAL	11-28-67 / 1 GAL	1-6 / 1 GAL					12 gal.
___	Hobert's Lima	10-28-67 / 12 GAL	2-4-68 / 1 GAL								12 gal.
___	Osgood Dragline	10-28-67 / OK									4 gal.
___	Marion Shovel	10-28-67 / 7 GAL	11-24 / 1 GAL	2-12 / 1 GAL	3-6-69 / 1 GAL						7 gal.
___	Northwest Shovel	10-28-67 / 7 GAL	11-30-69 / 2 GAL, 12-1-69 / 2 GAL	2-2-69 / 2 GAL	5-22 / 1 GAL						7 gal.
___	D-7 Caterpillar	10-28-67 / 5 GAL									5 gal.
___	D-8 Caterpillar(old)	10-28-67 / 6 GAL	12-22-19 / 2 GAL, 1-6-69 / 1 GAL								6 gal.
___	D-8 Caterpillar (Jesse)	10-28-67 / 6 GAL									6 gal.
___	TD-24 (Red)	10-28-67 / 7 GAL	12-26-69 / 3 GAL, 1-1-68 / 2 GAL	2-14-69 / 2 GAL	3-22 / 1 GAL						7 gal.
___	TD-24 (Yellow)	10-28-67 / DRAINED									7 gal.
___	Air Compressor	AT SHELBY									
___	Lincoln Welder	10-28-67 / 1 GAL									1 gal.
___	Fuel truck	10-28-67 / 1½ GAL	12-8 / ½ GAL, 2-24								1½ gal.
___	Cement mixer (drained)										
___	TD-14	10-28-67 / 5 GAL	1-19-68 / 1 GAL								5 gal.
✓	Half-track	10-27-67 / 1½ GAL									1½ gal.
___	Drill	10-28-67 / 1½ GAL									1½ gal.
___	Euclid #13	10-28-67 / 5 GAL	11-9-69 / 1 GAL, 12-8 / ½ GAL	2-1-68 / 1 GAL, X-8-69	3-6 68 / 1 GAL		#12	2-13 / 1 GAL, 2-24 / 3 GAL			5 gal.
___	Euclid #18	10-28-67 / 5 GAL	11-15-17 / 1 GAL								5 gal.
___	Euclid #28	10-28-67 / 6 GAL	11-27-67 / 1 GAL, 1-6-68 / 1 GAL	2-8-68 / 2-22-69 / 1 GAL							6 gal.
___	Euclid #29	10-28-67 / 6 GAL	11-14-69 / 1 GAL, 12-8-69 / 1 GAL	2-2-69 / 1-16-68 / 1 GAL	2-5-68 / 1 GAL, 4-28 / 12 GAL						6 gal.
___	Euclid #30	10-28-67 / 6 GAL	11-18-69 / 1 GAL, 12-28-69 / 1 GAL	2-12-69 / 1 GAL							6 gal.
___	Euclid #31	10-28-67 / 6 GAL	11-19-19 / 1 GAL, 1-19-69 / 1 GAL	2-19 / 1 GAL	3-22 / 1 GAL						6 gal.
___	54 3/4 T. Ford Pickup (drained)										
___	56 3/4 T. Ford Pickup	10-28-67 / 1½ GAL									1½ gal.

The Champion Mines

EQUIPMENT SERVICE CHECKLIST P. 2

Check When Serviced		Anti-Freeze added later										
✓ 57 3/4 T. Pickup (ford)	10-29-69 1½ GAL											1½ gal.
___ 59 ½ T. Ford Pickup	10-28-69 1½ GAL											1½ gal.
___ 64 Chev. Stationwagon	10-28-69 1½ GAL											1½ gal.
___ 10" Jaegar Diesel Pump	DRAINED											3 gal.
___ 6" Jaegar Diesel Pump	DRAINED	1-13-69 1 GAL										
___ 6" Jaegar Gas Pump	DRAINED	1-5-69 1½ GAL										
___ 4" Jaegar Gas pump	DRAINED											
✓ Motor Grader	10-28-69 3 GAL											4 gal.
✓ F-600 Ford	10-27-69 1½ GAL											1½ gal.
✓ Chev. Water truck	10-27-69 1½ GAL											1½ gal.
___ GMC dump truck (drained)												
___ Power Plant (Blue)												
___ Power Plant (Red)												
___ F-8 Ford Dump truck - PARKED												
___ F-8 Yellow drill truck	10-28-69 1½ GAL											1½ gal.
___ F-8 (Jack's Red Dump truck)	10-28-69 ½	1-4-69 1 GAL										1½ gal.
✓ 63 Light Blue Pickup	10-27-69 1½ GAL											1½ gal

The Champion Mines

Shipping Ticket

CHAMPION BROWN ORE

L. & N. No._____   Shipped_____ 194__

To_____

_____ Furnaces

From SHOOK & FLETCHER SUPPLY CO.   Birmingham, Ala.
 ORE DEPARTMENT

Timesheet

The Champion Mines

TIME, WEEK ENDING 8-1-1956										
NAMES	S	M	T	W	T	F	S	Total Time	Rate	Amount
Arthur Fulenwider	6	6	6		6	6	6			
Lonnie Cornelius	6	6	6		6	6	1			
Oscar Walker	6	6	6		6	6	6			
Hubert Henderson	6	6	6		6	6	6			
C. B. Cornelius	6	6	6		6	6	6			
Cecil Shade	6	6	6		6	6	6			
Jack Clements	X	6	6		6	6	6			
B. A. Sherman	6	6	6		6	6	6			
Jessie Henderson	6	6	X		6	6	6			
Oscar Hathcock	6	6	6		6	6	6			
L. M. Brothers	6	6	6		6	6	6			
Jessie Fulenwider	6	6	6		6	6	6			
H. V. Morton	6	6	6		6	6	6			
Herbert Hargus	6	6	2		6	6	6			
Jim Hickie	6	6	6		6	6	6			
Tom Green	6	6	6		6	6	6			
E. H. Clements	6	6	6		6	8	7			
Clarence Davenport	6	6	6		6	6	6			
Lossie Byrd	6	6	6		6	6	6			
Walt Williams	6	6	6		6	6	6			
Fred Cornelius	6	6	6		6	6	6			
Arthur D. Madden	6	6	6		6	1	1			

TIME, WEEK ENDING 8-8-1956										
NAMES	S	M	T	W	T	F	S	Total Time	Rate	Amount
		6	6	6		6	6	6		
		6	6	6		6	6	6		
		6	6	6		6	6	6		
		6	6	6		6	6	6		
		6	6	6		6	6	6		
		6	6	6		6	6	6		
		6	6	6		6	6	6		
		6	6	X		6	6	6		
		6	6	X		6	6	6		
		6	6	6		6	6	6		
		6	6	6		6	6	6		
		6	6	6		6	6	6		
		6	6	6		6	6	6		
		6	6	6		6	6	6		
		6	6	6		6	6	6		
		6	6	6		6	8	6		
		6	6	6		6	6	6		
		6	6	6		6	6	6		
		6	6	6		6	6	6		
		6		6		6	6	6		
		6	6	6		6	6	6		
		6	6	6		6	6	6		

About the Writers and the Blount County Memorial Museum

ABOUT THE WRITERS AND THE MUSEUM

AULDEN WOODARD

Born in Tait's Gap and reared in Champion and Oneonta, **Aulden Woodard**'s father and grandfather were miners. Woodard was the paperboy and the community messenger for Champion.

Prior to retiring as a Program Analyst from the Navy's Polaris Missile Facility Atlantic in Charleston, South Carolina, he enjoyed an impressive thirty-five year career in rockets, beginning with this country's early space program.

At age 26, he joined Dr. Wernher von Braun's rocket team and worked as a space and rocket engineer for the US Army and the National Aeronautics and Space Administration (NASA), in Huntsville, Alabama. As part of the early space program, well before launches to the moon and space walks, he played a key role in America's dream to place a man on the moon in 1969. His engineering assignments included Explosives, Rocket Engines, Cape Launch Coordinator and the SkyLab Environmental System for the Astronauts. Woodard now lives with his family in Goose Creek, South Carolina.

Van Gunter

Born in Talladega to Howard C. and Edna Mae Vandegrift Gunter in 1941, Van Gunter moved to a new mining camp at Doc Ray brown ore mine near Woodstock when his father, Howard Gunter, accepted a job as mine engineer and assistant superintendent under father-in-law E.N. Vandegrift, superintendent. After spending the WWII years with his young family in the Army at Neosha, Missouri, and Orlando, Florida, Howard Gunter returned to his job at the Shook & Fletcher Doc Ray mine in 1946. Edna Gunter grew up in the Taits Gap mining camp from 1920-1937; she was content with returning to the slow-paced life at Doc Ray after five years of Army life.

Van attended public schools at nearby Greeley Elementary and Brookwood High and graduated from the University of Alabama in 1963 with a BS in Mechanical Engineering. While in college Van worked two summers at the Champion Mine learning first-hand the hard physical work of ore mining when in 1961 he helped close and dismantle the Taits Gap mine washer and move it to the new site at Champion. He is a third generation brown ore miner with both grandfathers and father having been ore miners. This publication is a long awaited goal of his, having inherited mining records and diaries from his grandfather Ed Vandegrift and absorbed the history through living in a mining camp like Tait's Gap and Champion.

After graduation, Van was employed with Teledyne Brown Engineering, Huntsville, as a contractor at NASA Marshall Space Center. Directed by Dr. Werner von Braun, he work as a propulsion engineer on the Saturn I and Saturn V rocket booster launch vehicles for the Apollo Space Program, which successfully landed men first on the moon in August 1969 and returned them safely to earth.

In 1971 Van moved to New Orleans' Michoud Space Facility with Chrysler Space Division on the NASA Skylab Project which was successfully launched in orbit and manned by the Apollo capsule launched by Chrysler built Saturn S-IB booster rocket.

After work was completed on the Skylab launch project, Van worked on the NASA Space Shuttle propulsion system design at Chrysler Michoud for bid on the NASA contract won by Martin Aerospace to build the disposable external fuel tank. The last Space Shuttle flight was in 2011.

In 1974, Van returned to his native Alabama, working for 19 years with Southern Company Services as a fuels engineer in the purchase and utilization of coal for the 26 coal-fired power plants in the four state area of the Southern Electric System. There he inspected mine and plant facilities in all the major coal producing states in the nation, standardizing the sampling and testing methods to comply with purchase agreements.

After retiring from Southern Services, Van focused on historical interests with the restoration of the remaining buildings at Champion Mine from the 1960's final mining era, where he hosted a quarterly meeting of the Blount County Historical Society, of which he is a charter member. He is an active member in the Sons of Confederate Veterans and serves as a director of the John W. Inzer Museum in Ashville, owned and operated by the Heritage Organization.

Van lives in Hoover, Alabama and is married to the former Sara Rampy. They have two sons, Evan and Justin, and enjoy their three grandchildren: Joshua, Eva Mae, and Jonah. He and his sister, Jane Gunter Doughty, together own the former Champion mine property covering over 600 acres of the original pits, lakes and timber. Van and family enjoy their cabin retreat on the property where he collects mining memorabilia at his workshop barn. His previous coal testing experience prompted Van to sample and test the remaining iron oxide minerals - limonite and goethite, for suitability as pigments for coloring concrete and brick products, paints, and plastics; if the resulting evaluations are favorable, the history of mining at Champion may have another chapter in the 21st century.

AMY RHUDY

Amy Rhudy, Curator
Blount County Memorial Museum

Amy Rhudy, dedicated to preserving the county's history, worked first as a part-time employee at the **Blount County Memorial Museum** before being appointed the Museum Curator in 1999. She is married to Leland T. Rhudy; they have a son Nathan and live in Oneonta, Alabama. The Museum opened to the public in 1970. It continuously updates and maintains an array of historical artifacts and documents on the history of this county and its people.

One museum section includes the history of mining and mineral deposits in Blount County. The Museum has received a large display of the **Champion Mines**, with photographs, maps, documents, and artifacts from those days of yore, donated by **Van Gunter**, historian and current owner of the property where the mine was located. The display, contains scores of ore samples and other items connected to the iron ore industry.

Gunter held a reunion with the miners to reminisce about their hardships, antics, and camaraderie at the mine. Gunter organized **the Champion Miners Society** which asked the Museum to prepare a book of collected materials in appreciation of their heritage.

The Museum has assembled this book from selected archived items with the help of **Aulden Woodard**, former Champion resident and paperboy, and many other miners. It has stories of the lives and times of Champion and Tait's Gap miners, their school, church, history, residents' maps, Miners Honor Rolls, articles, memorabilia, and photos.

INDEX

Accident 80, 149
Adkins 113, 121, 149, 156, 157, 158, 159, 160, 162, 164, 169, 196
Adkins Mine 113, 149, 156, 157, 158, 159, 160, 162, 164, 196
Alabama Museum of Natural History .. 150
Alabama Power Company 18, 51, 161 107, 108
Alexander, John 81
Alice................................. 97, 113
Altoona 73, 77, 78, 79, 85, 93, 94, 95, 98, 99, 125, 168
Appalachian.................. 11, 22, 36, 70
Armstrong 21, 81, 100, 107, 108, 112
Armstrong, James 81
Auditor 110
Auxford Mine-Russellville............... 150
Baseball 30, 61, 63
Batch................................. 112
Battles, Cliff 81
Beasley....................... 81, 100, 112
Beason 81, 100, 112
Bemiston 107, 112
Bird Lane 71, 72
Bird School 37
Birmingham Mineral Railroad 16, 40, 41, 91, 93, 94
Birmingham water 17
Brasseal................................ 147
Brown ore 107, 108
Buckner........... 81, 111, 112, 133, 134
Bynum 37, 39, 74, 81, 94, 100, 107, 108, 112, 116, 121, 123, 138, 143, 160, 162
Byrd 81, 100, 110, 112, 143, 152, 154, 156, 165
Caffee............. 156, 157, 159, 162, 164 1, 4, 5, 11, 12, 14, 16, 17, 18, 19, 20, 21, 22, 23, 24, 25, 30, 31, 32, 33, 34, 35, 36, 37, 38, 39, 40, 41, 46, 47, 49, 50, 51, 52, 53, 54, 55, 57, 58, 60, 61, 62, 63, 64, 65, 68, 69, 72, 73, 74, 80, 81, 84, 85, 86, 88, 89, 91, 92, 93, 94, 95, 97, 98, 99, 107, 108, 111, 112, 113, 114, 115, 116, 117, 118, 119, 120, 121, 122, 123, 124, 126, 129, 130, 131, 132, 133, 134, 135, 136, 137, 138, 139, 140, 141, 142, 144, 146, 147, 148, 151, 157, 158, 159, 160, 161, 162, 163, 164, 166, 167, 168, 169, 173, 187, 195, 196, 198, 208, 211
169, 173, 211
CHAMPION MINERS HONOR ROLL.. 28
Champion Road 12, 22, 23, 36, 51, 57, 61, 148, 163
Champion Spring........................ 148
Chandler 39, 81, 107, 108, 111
Chandler, Oscar........................ 81
Chepultepec 21, 27, 41, 92
Church 23, 52, 53, 57, 71, 72, 73, 74, 98, 99, 100, 110, 117, 119, 120, 122, 124, 125, 132, 133, 148, 156
Civil War................................ 16
Clements, Jack 81
Clements, Jess 81, 152
Cleveland 63, 77, 79, 92, 96, 127
Cold Water Creek 148
Commissary 20, 21, 23, 25, 34, 35, 36, 49, 61
Compton Mines......................... 26
Confederate.................. 86, 112, 148
Convict 36
Cooperative Extension Service 69
Cornelius, Fred 81
Cornelius, Lonnie....................... 81
Courthouse 93
Cummins diesel 158
Daily, Will............................... 81
Dam 17, 55, 111, 118, 120, 121, 122, 123, 124
Depression 24, 33, 50, 68, 70, 72, 77, 125, 129, 131, 133
Diary................................... 168
Diesel.................................. 151
Dinkey 18, 54, 72, 80, 109, 120, 125, 143, 162, 198
Doc Ray 112, 113, 121, 160, 196
Doud............................. 132, 133
E. I. Dupont 109
E. N. Vandegrift 17, 20, 41, 47, 65, 72, 98, 100, 102, 111, 112, 115, 131, 133, 152, 154, 157, 159, 164, 168, 196
Electrical 123
Euclid 18, 149, 156, 157, 158, 159, 160, 161, 163, 164, 165
Eureka Mining Co. 87
farmer 57, 98, 131, 133, 139
Fendley-Hagood Co..................... 109

212

First Baptist Church 37
Frank C. Cheney 92
Furnace 17, 19, 88, 93, 107, 151
Galbeath 30
Galbreath 81, 100, 112, 143, 160
Galbreath, Webster 81
Galbreath, Will 81
George 15, 34, 35, 82, 83, 85, 96, 108, 168
Graystone 41
Green, Tom 81
Gulf States Steel 63, 97
Gunter 4, 49, 55, 74, 81, 84, 98, 107, 112, 115, 146, 148, 157, 162, 163, 164, 168, 211
Gypsum Crystal 150
Halloween 36, 37
Hanby 14, 15, 16, 19, 20, 65, 85, 86, 89, 168
Hartley, Thomas 81
Hay 121, 124, 153
H-Bomb Plant 149
Heavy Media 151
Hill 18, 23, 36, 59, 82, 100, 107, 108, 112, 116, 145
Hill, J. C. 82
History of Blount County 5, 15, 168
Home Demonstration Club .. 5, 32, 50, 69
Homestead Works 97
house 23, 30, 33, 34, 35, 37, 49, 50, 53, 57, 68, 71, 74, 93, 98, 100, 101, 102, 110, 112, 113, 114, 118, 119, 124, 125, 126, 138, 139, 147, 149, 157, 160, 162
Huie 38, 74, 82, 100, 112
Hunting 115, 116, 117, 126, 127
Insurance 109
Iron Making in Alabama 26
Jones 49, 50, 82, 100, 116, 119, 120, 122, 130, 133, 134, 136, 138, 150
Kelton 100, 112, 121
Know Your America Program 17
L & N Railroad 16
Lawrence 21, 54
Layout of Houses 31
Lester Memorial Methodist Church 37, 119
Limonite 107
Lowry, Elton 82
Lybrand 4, 50, 80, 82, 100, 107, 108, 112, 136, 137, 143

Lybrand, Quinton 82
Lybrand, Sam 82
Lybrand, W. W. 100
Map 92, 148, 169, 187
Marion 1931 Traction 147
Marker 12, 20
Mary Pratt Furnace 90
Million Dollar Deal 89
Morrison 51, 55, 82, 100, 110, 112, 135, 138, 140, 143, 150, 152, 154, 156
Movie 36
muck 18
Mule 18
80, 112
Odenville 53, 54, 64, 80, 98, 136, 141, 150, 196
1, 5, 12, 15, 16, 17, 18, 20, 21, 23, 25, 26, 29, 30, 32, 33, 35, 36, 37, 38, 39, 40, 41, 42, 46, 47, 50, 51, 57, 61, 63, 64, 72, 73, 74, 76, 79, 85, 86, 89, 91, 92, 93, 98, 109, 110, 115, 116, 117, 119, 121, 122, 123, 124, 125, 126, 127, 130, 131, 133, 134, 135, 137, 138, 139, 140, 148, 163, 169, 174, 208, 211
Oneonta Grammar School 37
Oneonta Ore & Mining 21
Oneonta, New York 16, 91
Onto 92
Ore Mining Terms 166
Oxmoor Furnace 87
Park 5, 41, 46, 61, 63, 92
Payroll 152, 154, 156
Pinson 14, 85, 86
Pittsburgh 96
ponds 25, 50, 153, 155, 157, 160, 161, 162, 164, 167
Price 82, 112, 113, 151, 195
Pritchard 88
Radio Hill 162
Railroad Incorporation Act of Alabama . 16
Recreation 61
Red Cross 131, 133, 134
Red Hill 35, 36, 49, 61, 148
Red Mountain 17, 20, 21, 22, 23, 26, 46, 77
Republic 20, 65, 94, 139
Robbins Coal Company 62
Rousseau 86, 87, 148
Rush 82, 100, 112

213

S&F 96, 97, 98, 103, 104, 107, 108, 109, 110, 112, 113, 116, 117, 118, 119, 120, 121, 122, 130, 131, 133, 135, 144, 146, 147, 148, 150, 151, 155, 159, 162, 167, 196
Sand Valley 22
School 21, 30, 35, 36, 37, 38, 39, 76, 89, 92, 96, 115, 120, 148, 156
Shantle process 88
Shipments 153, 155, 156
Shook 17, 18, 20, 41, 47, 49, 53, 54, 55, 61, 65, 72, 80, 93, 96, 97, 98, 102, 107, 108, 109, 110, 111, 113, 114, 115, 116, 117, 118, 119, 120, 121, 122, 123, 124, 125, 126, 127, 128, 129, 130, 131, 132, 133, 134, 135, 136, 137, 138, 139, 140, 141, 142, 144, 145, 146, 147, 149, 150, 151, 157, 159, 162, 164, 168, 169
Shook and Fletcher Mines 18, 47
Shook and Fletcher Supply Company .. 41
Sloss 16, 17, 18, 19, 20, 47, 65, 88, 89, 90, 93, 98, 109, 110, 114, 116, 118, 124, 125, 151, 157, 164, 169
Sloss-Sheffield Steel & Iron 98, 151
Spout Spring Gap 22
St. Louis 96
Steam shovel 62
Story of Coal and Iron in Alabama 15, 16, 168
Straight Mountain 17, 22, 24, 25, 29, 47, 49, 52, 53, 54, 64, 73, 74, 76, 86, 87, 99, 108, 113, 147, 148, 149, 160
Survey 85, 94, 107, 169, 190, 195
Susan Moore............................... 92
Syrup 57, 58
Syrup Festival......................... 57, 58

T.C.I. 20, 24, 49, 53, 54, 61, 65, 85, 89, 93, 94, 95, 96, 97, 114, 116, 118, 119, 121, 125, 126, 129, 130, 131, 133, 134, 136, 137, 144, 145, 147, 148, 151, 157, 164
Tailings 153, 155, 156
Talladega 18, 47, 60, 61, 74, 86, 98, 99, 107, 112, 136, 139, 140, 148
TCI 137, 144
Teachers 35, 38, 39, 76, 102
Tennessee Coal and Iron Railroad 21
Tennessee Company 85, 89, 94, 96, 97, 145, 149
The Birmingham Post 20, 32, 50, 69
The Heritage of Blount County 16, 91
The Southern Democrat 17, 20, 21, 47, 63, 73, 74, 132, 133, 195
Thomas Worthington Company. 112, 115
Tonnage 195, 196
Tornado 110
Vandegrift 5, 14, 17, 18, 19, 24, 46, 47, 48, 51, 54, 55, 56, 68, 69, 83, 84, 98, 99, 100, 101, 102, 107, 108, 109, 110, 111, 112, 113, 114, 115, 116, 119, 129, 130, 131, 132, 133, 136, 143, 147, 148, 152, 157, 159
W. M. Morton 112
Wall Street Crash 129
Washer 23, 24, 49, 50, 61, 129, 143
Water 14, 94, 118, 148, 157, 161
Williams, Walt............................. 83
Woodard, Grady.......................... 83
Woodstock 25, 96, 112, 113, 121, 156, 169

The Champion Mines